Synthetic Aperture Radar Systems

Theory and Design

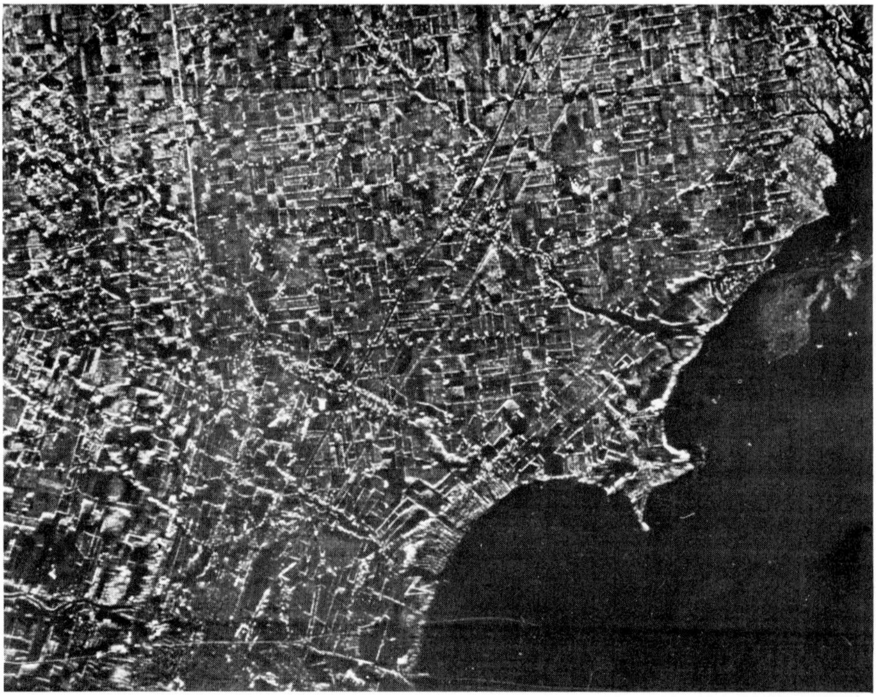

Synthetic aperture radar image of Monroe, Michigan area.

Synthetic Aperture Radar Systems

Theory and Design

Robert O. Harger

DEPARTMENT OF ELECTRICAL ENGINEERING
UNIVERSITY OF MARYLAND
COLLEGE PARK, MARYLAND

formerly

RADAR AND OPTICS LABORATORY AND
DEPARTMENT OF ELECTRICAL ENGINEERING
UNIVERSITY OF MICHIGAN
ANN ARBOR, MICHIGAN

1970

ACADEMIC PRESS New York and London

ACADEMIC PRESS, INC.
111 Fifth Avenue, New York, New York 10003

United Kingdom Edition published by
ACADEMIC PRESS, INC. (LONDON) LTD.
Berkeley Square House, London W1X 6BA

LIBRARY OF CONGRESS CATALOG CARD NUMBER: 70-91430

PRINTED IN THE UNITED STATES OF AMERICA

ELECTRICAL SCIENCE
A Series of Monographs and Texts

Editors: Henry G. Booker and Nicholas DeClaris
A complete list of titles in this series appears at the end of this volume

To Janet
and Eric, Peter, Karl, and Laurie

Preface

The synthetic aperture radar system is perhaps the most important member of the recently emerging class of radar systems whose primary purpose is to form an image of very large objects such as areas of terrain. It is also a member of the class of remote sensors that make use of the detailed structure of the scattered signal by means of rather complex data processing. These two qualities of synthetic aperture radar render its theory and practice quite distinct, in important respects, from the well-known theory and practice of "conventional" radar as summarized by, for example, Berkowitz (1965).

After fifteen years of research, development, and application, the theory and practice of the synthetic aperture radar system is well established and ready to take its place as an important and powerful remote sensing technique, with the possibility of even more interesting applications to diverse areas of science.

The purpose of this book is to summarize the principles, theory, and system design of the synthetic aperture radar both to serve as a guide for system considerations and as a basis for further research and development of the system, especially as new requirements emerge from new areas of application.

The language of systems studies is mathematical, and this dictates the style of most of the discussion. We begin by discussing the principles of the system as simply and with as little mathematics as possible, however.

The main theme of the book is the application of the ideas of communication theory to this system. In fact, we give rather explicitly a discussion of signals and errors and what to do about them, that proceeds well beyond the treatment in communication theory texts. It is hoped that a good deal of the discussion will find application to other systems that process data in an analog manner. Specifically, the analysis and design of systems when phase errors, ambiguities, motion errors, and nonlinearities are present and when the scattering mechanism is important are discussed in detail.

The book is written on three levels. The first two chapters can be read with understanding by almost anyone with any interest in the subject. These two chapters, along with the simplified signal-to-noise ratio calculation of Sec. 3.6, give an easy introduction to the principles, important parameters, and elementary system design of synthetic aperture radar systems.

Chapters III–VIII discuss the optimum design of these systems and the effect of various kinds of errors that are peculiarly important in systems of this type. This level would be of most interest to those concerned with the engineering and system design of synthetic aperture radar. The optimality criteria are concerned with resolution, additive noise, and ambiguity. The effects of phase errors, nonlinearities, motion errors, and the scattering mechanism on such systems are considered.

Lastly we consider the design of these systems, recognizing at the outset that ambiguities or phase errors may be the most important consideration. The resulting systems are even more complicated, yet some of them are being used in synthetic aperture radar systems. This level would be of interest to those performing research and development on these systems.

An outline of topics discussed is given in Sec. 1.6.

This book is based in part on lectures prepared for an intensive course offered at the University of Michigan annually since the summer of 1966; the other lecturers and students undoubtedly influenced the exposition. The author has also benefited greatly from discussions with other members of the Radar and Optics Laboratory of the University of Michigan. A particular debt of gratitude is due to Professor W. M. Brown and Dr. C. J. Palermo for numberless discussions; and to Professor E. N. Leith, whose wonderful insight has contributed so greatly to the understanding and success of synthetic aperture radar. Helpful comments were made on the manuscript by Drs. C. J. Palermo and J. S. Zelenka, and on Chapter V by Professor E. N. Leith.

ROBERT O. HARGER

Ann Arbor, Michigan
November 1969

Contents

PREFACE vii
GLOSSARY OF SYMBOLS xii

I. Introduction

1.1 Introduction 1
1.2 An Elementary Terrain Imaging Radar 3
1.3 The Elementary Synthetic Aperture Terrain Imaging Radar 7
1.4 The Elementary Optical Processor 9
1.5 Terrain Imaging Radar Applications 13
1.6 Topics to Be Discussed 14
Comments and References 17

II. Synthetic Aperture Principle and Radar: Simplified Discussion

2.1 Introduction 18
2.2 The Synthetic Aperture Principle 19
2.3 Synthetic Aperture Radar 26
2.4 The SAR Doppler Space 30
2.5 The Azimuth Modulation 37
2.6 The Linear Frequency Modulation 43
2.7 Sampling Effects 49
2.8 Digital Processing 53
Appendix 2A: Azimuth Modulation Spectrum 55
Comments and References 58

III. SAR Signal, Image, and SNR

3.1 Introduction 59
3.2 The Received Signal 60
3.3 Signal Spectrum: Point Scatterer 66
3.4 Matched Filtering: Point Scatterer 68
3.5 Linear Characterization of SAR 70
3.6 Signal-to-Noise Ratio 75
Appendix 3A: Reflectivity Density Calculation by the Method of Physical Optics 79
Comments and References 82

IV. System Design A

4.1 SAR System Models 83
4.2 Criteria for No Noise 86
4.3 Criteria for Additive Noise 92
4.4 Comparison of Criteria 102
Appendix 4A: Spectral Density of Two-Dimensional System Noise 103
Appendix 4B: Image Calculation by Method of Geometrical Optics 104
Comments and References 108

V. Optical Processing

5.1 Elements of Optical Processing 109
5.2 Optical Compression 121
5.3 LIV Filtering 126

VI. Phase and Motion Errors

Part I: Phase Errors

6.1 Introduction 129
6.2 Effects on Phase-Matched System 131
6.3 Effects on LMVE System 141
6.4 Effects on Ambiguity Function 144
6.5 Effect of Dispersive Errors 145
6.6 Bounds on Phase Errors for System Design 145

Part II: Motion Errors

6.7 Introduction 146
6.8 Large Bandwidth Motion Errors 149
6.9 Small Bandwidth Motion Errors 151
Appendix 6A: Calculation of Some Expectations 154
Comments and References 154

VII. Nonlinearities

7.1 Introduction 156
7.2 The Transform Method 156
7.3 Multiple Signals and Noise: Hard Limiting 160
7.4 Random Signals and Random Noise 166
Comments and References 171

VIII. Optimum Ambiguity Design

8.1 Introduction 172
8.2 Ambiguity Function Design Criteria 173
8.3 Solution of the Antenna Design Problem 176
8.4 Solution of the Signal Design Problem 185
8.5 Example 188
Appendix 8A: Alternate Criterion with Azimuth Offset 190
Appendix 8B: The Phase Modulation That Relates $|F|$ and $|\tilde{F}|$ 192
Appendix 8C: Solution of Nonlinear Design Problem for Matched Filtering 194
Comments and References 196

IX. The Ambiguity Function

9.1 Introduction 197
9.2 General Properties 200
9.3 The Range Extension Problem 207
Appendix 9A: A Defective Ambiguity "Solution" 213
Comments and References 214

X. System Design B: Phase Errors

10.1 Introduction 215
10.2 Resolution and SNR_{max} Criteria 215
10.3 LMVE with Phase Errors 220
10.4 Maximum Likelihood Criteria 223
Comments and References 231

References 233

Index 237

Glossary of Symbols

The following is a list of symbols and abbreviations that appear at several points of the discussion. The tilde (e.g., $\tilde{F}$) denotes Fourier transform (e.g., F) throughout.

A	(i) aperture area
	(ii) set of points (x, z) belonging to aperture
$A(t)$	$B_h(x)$, $x \equiv vt$
B_h	azimuth complex modulation [Eq.(3.8a)]
B_X	processed X-length segment of B
B_n	processed X-length segment of nth alias of B
B_v	complex vertical antenna pattern [Eq.(3.8b)]
$\not{C}$	arbitrary undefined complex constant
D_h	physical aperture "horizontal" extent
D_v	physical aperture "vertical" extent
E	scalar E/M field
E_{0h}	physical aperture "horizontal" illumination function
E_{0v}	physical aperture "vertical" illumination function
$E\{\cdot\}$	statistical expectation (Chapter VI)
f	transmitted wave form
F	complex modulation of radiated wave form of SAR system
F_m	$\|F\|$, real envelope
FM	frequency modulation
f_s	sampling frequency
g	reflectivity density ψ referred to "terrain plane"
G	complex modulation of g
$\hat{g}$	SAR image of g
$\hat{G}$	complex modulation of $\hat{g}$
G_1	complex reflectivity density of specific object
G_c	reflectivity density of clutter
h	receiver and/or processor impulse response
h_r	recorder impulse response
H	complex modulation of h
I_h	azimuth beamwidth measure
I_v	"vertical" beamwidth measure
k	(i) wave number, $2\pi/\lambda$
	(ii) $4\pi v^2/\lambda R$, azimuth linear FM rate (Hz/sec)

LIV	linear, invariant (operator)
LQIV	linear, quasi-invariant (operator)
LMVE	linear, minimum variance estimation
ML	maximum likelihood
N	additive noise
N_1	additive noise due to RF noise
N_2	additive noise other than N_1
N_o	additive noise constant spectral density
P_{ave}	average radiated power, $\frac{1}{2}\|F\|^2 \cdot f_s$
P_n	spectral density of $\{N\}$
P_1	spectral density of $\{N_1\}$
P_2	spectral density of $\{N_2\}$
P_g	spectral density of $\{G\}$
P_c	spectral density of $\{G_c\}$
P_ε	spectral density of $\{\exp i\gamma\}$
P_γ	spectral density of $\{\gamma\}$
PRF	f_s
PSWF	prolate spheroidal wave function
$\|q\|^2$	ambiguity function of SAR [Eqs.(3.14) and (3.18)]
$\|q_1\|^2$	range channel ambiguity function
$\|q_2\|^2$	azimuth channel ambiguity function
R	slant range
R_g	ground range
ΔR	mapped slant range interval
ΔR_g	mapped ground range interval
SAR	synthetic aperture radar
SNR	signal-to-(additive) noise power ratio
$\text{SNR}_{\max}$	maximum signal-to-noise power ratio
T	temporal extent of processed azimuth modulation
T_S	temporal sampling period
TBP	time-bandwidth-product
u_o	azimuth offset frequency (rad/sec)
v	relative velocity of vehicle carrying SAR radar set
X	vT, spatial interval
X_s	vT_s
ε	motion error (Chapter VI)
γ_o	depression angle
γ	phase error (Chapter VI)
Ω	azimuth modulation bandwidth measure
Ω_h	$\frac{1}{2}D_h$
Ω_v	$\frac{1}{2}D_v$
Ω_s	half-bandwidth measure of F
φ	azimuth Doppler phase modulation
ψ	reflectivity density
ρ_a	azimuth resolution
ρ_τ	range delay resolution
σ	scattering cross-section density
ν	velocity of light
ω_c	RF carrier frequency (rad/sec)
ω_o	range offset frequency (rad/sec)

I

Introduction

1.1. INTRODUCTION

The synthetic aperture radar system was conceived of and is primarily used as a system for forming an *image* of terrain from a remote location. As are most radars, it is a scanning system; its rigidly prescribed scanning method provides its reason for existence, exemplifying that *a scanning system can have a narrow effective scanning beam without having a narrow radiated beam.*

The geometrical configuration of the synthetic aperture radar scanning method is shown in Fig. 1.1 in its most common arrangement with the radiated scanning beam orthogonal to the velocity vector of the vehicle, and at a depression angle with respect to the horizontal. Of course, the vehicle carries the necessary—and fairly conventional—electronic equipment for radiating the scanning beam and receiving the reflected waves. The scanning in the abeam, or range, direction is accomplished by emitting pulses of energy as is customarily done by radar systems; the scanning velocity in range is therefore that of light, $v \approx 3 \times 10^8$ m/sec. The scanning in the along-track, or azimuth, dimension is done by moving the vehicle; the scanning velocity in azimuth is therefore that of the vehicle, from about $\upsilon \approx 50$ m/sec (aircraft) to about $\upsilon \approx 8 \times 10^3$ m/sec (orbital vehicle). *The practical success of the synthetic aperture radar system* (hereafter denoted "SAR system") *depends upon this large disparity of scanning velocities, and the ratio* v/υ *is a fundamental SAR system parameter.*

The nature of the limitation imposed by the finiteness of the ratio v/υ is easily seen. If the SAR is to have a *system* scanning beam of the dimension shown in Fig. 1.1, then it is reasonable to assume that at least one pulse must be emitted during the time the vehicle traverses this distance. For as is well known in communication work, we cannot expect to be able to reconstruct an image of the terrain with detail or resolution finer than the sampling

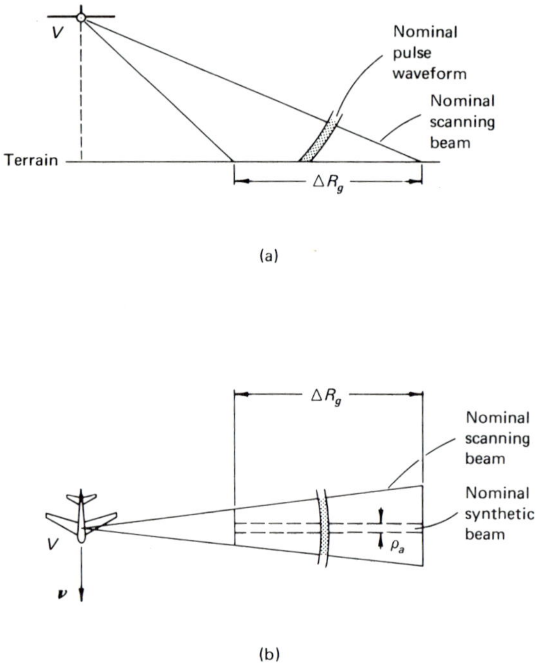

FIG. 1.1 Typical geometry of synthetic aperture radar system: (a) elevation; (b) plan.

interval, roughly speaking. Since the time required to sample a slant range interval is $2\,\Delta R/v$ sec, and the time required to traverse the resolution distance ρ_a m is ρ_a/v sec, equating these times gives the identity

$$\frac{2\,\Delta R}{\rho_a} = \frac{v}{v}. \tag{1.1}$$

In words, the slant range interval that can be mapped—measured in resolution units—is a constant inversely proportional to vehicle velocity. We shall, of course, spend some effort below to make this limitation more precise and investigate the possibilities of subverting it.

The fact that a scanning system can have an effectively narrow scanning beam even without a narrow radiated beam has, in fact, been long well known in radar work. Probably, the most familiar example is the linear frequency modulated ("FM"), or "chirp" signal. Such a signal pulse may have arbitrary time extent but its "fine structure," due to the frequency modulation, allows it to be reconstituted, or *compressed*, into a narrow time extent by a linear operation. As is also well known, a sequence of linear, time-invariant† operations can be arranged in arbitrary order without change in the composite linear operator. Thus, as long as all operations in a radar system, including those of transmission and reflection, remain linear, we can form the narrow pulse as the last operation just as well as if it were formed as the first operation.

The motion of the vehicle—and hence of the scanning beam—in the azimuth dimension causes just such a fine structure, due to phase modulation resulting from the Doppler effect. In fact, this phase modulation is well approximated in most SAR's by the linear FM, or "chirp" modulation.

The importance of this idea is simply that a narrow scanning beam requires a large antenna to form it, impractically large for air- or space-borne vehicles at desired or required remoteness. Thus, allowing a larger scanning beam enables use of a smaller, practical size antenna in important situations. As we shall see, the effective system scanning beamwidth in the azimuth dimension can be (1) of the order of azimuth dimension of the physical antenna's aperture, (2) independent of range and (3) independent of wavelength over a wide range of wavelengths. The last two attributes are, of course, not obtainable with a simple physical aperture.

1.2. AN ELEMENTARY TERRAIN IMAGING RADAR

In order to establish a common starting point for the discussion of synthetic aperture radar, we review the main features of elementary range measuring radar and show how it can be, and has been, used to form images of terrain.

For example, consider the rather basic radar range scanning system shown in Fig. 1.2a. The transmitter-receiver "set" can be of a conventional type familiar to most readers: the details are not important here, except that the transmitter generates a waveform with an amplitude modulation of short temporal duration; the radar's ability to distinguish—or resolve—closely spaced objects is determined by the length of the "waveform"—commonly termed a "pulse." The function of the antenna is, of course, to direct the radiated waveform onto the range interval that is to be imaged and to receive the waveform scattered, or reflected, from the terrain. The receiver basically converts the waveforms from a frequency band suitable for radiation, to a frequency band suitable for presentation via equipment, either permanently or not, to a human observer. A widely used presentation system is a cathode ray oscilloscope ("CRT"), which may be either visually observed or photographed. Of course, the CRT "spot" scans an interval on the tube face in synchronism with the scanning of the physical waveform pulse over the range interval: range is measured and indicated in this manner.

† We abbreviate "linear, invariant" by LIV.

The simplest possible useful model for such a system is shown in Fig. 1.2b. We assume a "reflectivity density, g," for the terrain and, as the reflection mechanism is generally linear, characterize the received signal as the "convolution, $g * f$," of g with the transmitted waveform f (right now we do not worry about amplitude scaling). The receiver can be characterized by an impulse response h so that the output of the radar *system* is

$$\hat{g} \equiv g * f * h. \tag{1.2}$$

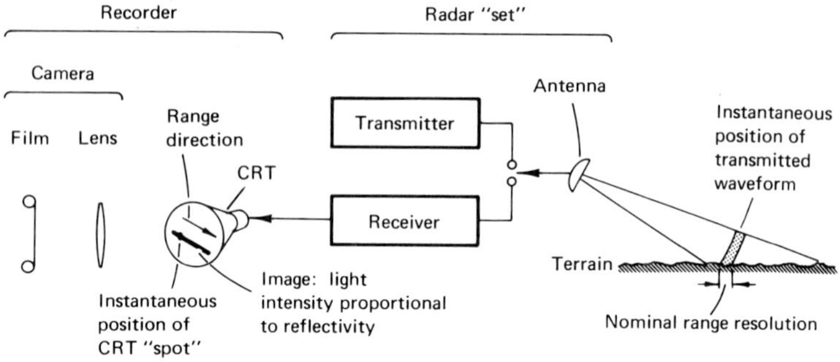

FIG. 1.2a An elementary range dimension imaging system.

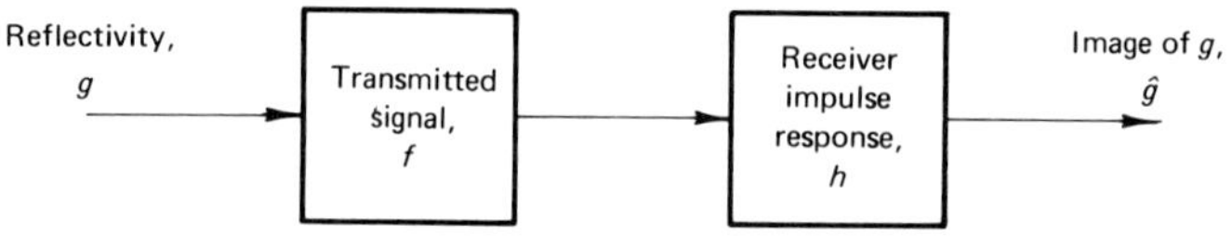

FIG. 1.2b The simplest useful analytical model for the range imaging system of Fig. 1.2a.

Of course, $\hat{g}$ *is an estimate of what* g *is*, or, to phrase it another way, the recording of $\hat{g}$ is our *image* of g.

We observe the mathematical triviality that we may have a special form for f, say $f = f_d * f_c$, without changing $\hat{g}$. But the physical consequence is anything but trivial. We may rewrite

$$\hat{g} = g * f_d * h_1 \tag{1.3}$$

if $h_1 = f_c * h$ and interpret it this way:

$$\hat{g} = \begin{Bmatrix} g, \text{ reflectivity} \\ \text{density} \end{Bmatrix} * \begin{Bmatrix} f_d, \text{ (new)} \\ \text{transmitted} \\ \text{signal} \end{Bmatrix} * \begin{Bmatrix} h_1, \text{ (new)} \\ \text{receiver} \\ \text{impulse} \\ \text{response} \end{Bmatrix}.$$

That is, *we can change the transmitted signal, as long as we can change the receiver, and leave the system response unchanged.* In particular, *a radar system can have resolution much smaller than the time duration of the transmitted pulse.* In radar practice, f_d is called a *dispersed pulse*, and f_c is called the *compression filter* for obvious reasons.

The recognition of this fact characterized in Eq. (1.3) was forced upon radar engineers by practical limitations in peak power that could be achieved by the radar transmitter and requirements for fine resolution, but equally fundamental was an assumption about the nature of spurious errors, or noise, with which the received signal might be perturbed (Chapters IV and VI).

A radar system that is modified to employ such a dispersed transmitted pulse is conceptually changed very little. The elementary terrain imaging system of Fig. 1.2 is shown in Fig. 1.3. The transmitter change can be much easier than the receiver change, which can be quite difficult. However, in Fig. 1.3a we have indicated that one could use the transmitter of Fig. 1.2a

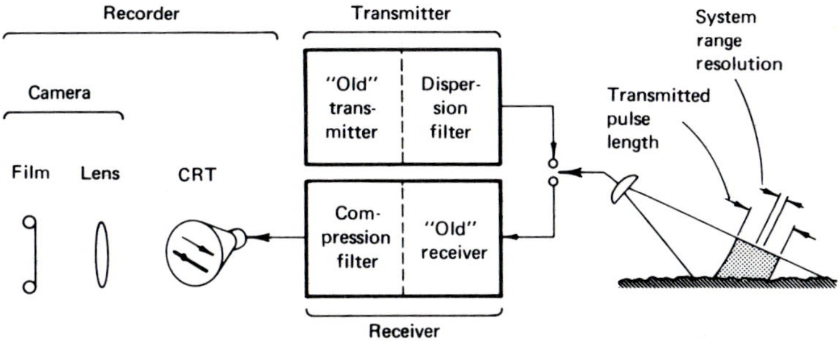

FIG. 1.3a An elementary range dimension imaging system employing signal dispersion and compression.

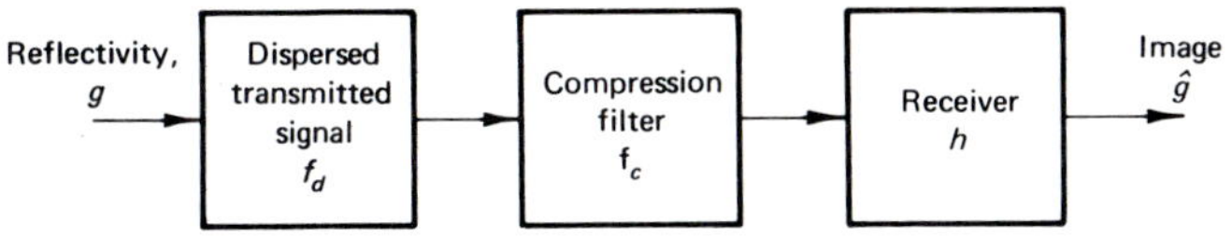

FIG. 1.3b The simplest useful analytical model for the range imaging system of Fig. 1.3a.

followed by a "dispersion filter" which conceptually is opposite in its effect to the compression filter. The final transmitter operation might then be a linear amplification to the desired power level. As we shall see below, and as is well known, the degree of difficulty of realizing the "compression" operation is roughly indicated by the "compression ratio," defined as the ratio of some measure of the width of the signal before and after compression. The required modification of the analytic model of Fig. 1.2b is shown in Fig. 1.3b.

It is, of course, easy to employ this basic one-dimensional—we are calling the dimension "range"—imaging system to make a two-dimensional, or terrain, imaging system. In practice it has been done two ways—and the two ways even combined. First, one could simply rotate the antenna beam, generally most simply physically rotating the antenna, to scan sequentially range radii. Second, one can physically move the radar by means of a vehicle, generally in uniform translational motion orthogonal to the range dimension, and periodically emit pulse waveforms. Such a system is shown in Fig. 1.4a. Just as the CRT spot is scanned in synchronism with the physically scanning radiated waveform, so the film translation is in synchronism with the advance of the vehicle. In this way range and azimuth are measured.

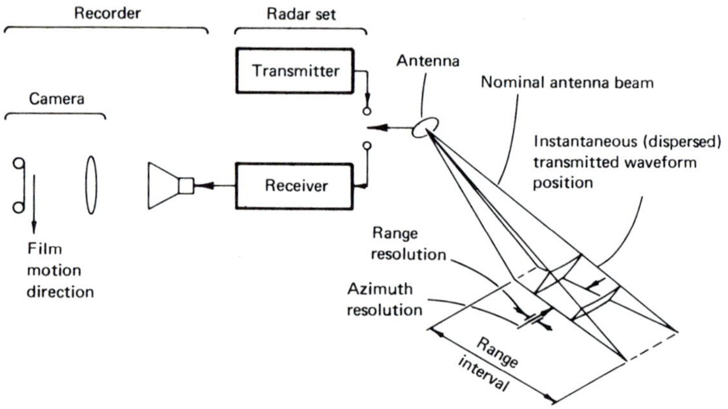

FIG. 1.4a An elementary terrain imaging radar system employing pulse dispersion in range. (The compression filter is included in the "receiver.")

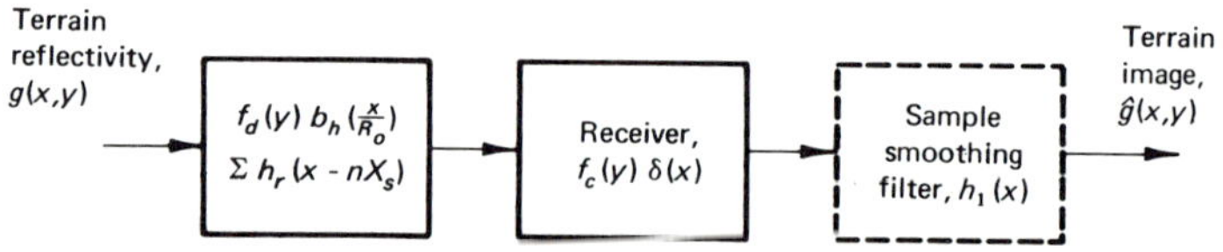

FIG. 1.4b An analytic model for the terrain imaging radar system of Fig. 1.4a. The blocks are LIV operations with indicated impulse responses. (The recorder impulse response can be chosen so that the impulse response of the first block is just $f_d(y)b_h(x/R_o)$ and the sample smoothing filter, h_1, is therefore omitted.)

Just as the range resolution, or fidelity, is achieved in the range coordinate by a narrow pulse, resolution, or fidelity, is achieved in the orthogonal or "azimuth" dimension by using a physically narrow beam. It is obvious from the geometry that azimuth resolution is a function of range implying, in particular, that azimuth resolution is not constant over the mapped range segment, and that azimuth resolution is degraded by increasing remoteness of the vehicle-borne radar set from the range interval to be mapped. It is also quite well known that the azimuth resolution will be a function of the radiated (mean) wavelength.

A simple useful model that can be abstracted from this system is shown in Fig. 1.4b. It is not obvious, but reasonable that, again with a linear reflection mechanism, the vehicle motion results in a convolution of the azimuth antenna pattern, b_h, with the reflectivity g in the azimuth direction—say coordinate x—just as the motion of the propagating pulse resulted in a convolution of the transmitted signal f_d with g in the range direction—say coordinate y And of course this scanning is done at a spacing X_s commensurate with the desired azimuth resolution. Hence the nature of the impulse response of the first block of Fig. 1.4b. The two-dimensional data must be compressed in the range, or y dimension, and this is accomplished by the impulse response factor f_c in the second box; the radar set's receiver normally has no effect in the azimuth direction—hence, the other factor $\delta(x)$. At the output of this second filter the data is, in our model, still effectively sampled in azimuth, and it must be restored, or "smoothed," by a filter $h_2(x, y)$ which may act effectively ideally in the range (y) direction. In practice h performs the sample smoothing and can be realized in the recorder by proper selection of recording film—which is to a first order approximated by a (two-dimensional) low pass filter—and film velocity, as we shall discuss at some length in Chapter III.

1.3. THE ELEMENTARY SYNTHETIC APERTURE TERRAIN IMAGING RADAR

We can now outline a radar system that can provide a *system* azimuth resolution better than that due to the physical antenna's beamwidth. The motion of the vehicle bearing the radar over the terrain causes a Doppler phase modulation that gives the two-dimensional signal a structure that will compress in the azimuth dimension. This latter fact is easy to establish and, in fact, is where we will begin an analysis of the SAR in Chapter II.

Therefore, referring to the imaging system of Fig. 1.4a, we now must add a final processing operation that compresses the data in the azimuth direction (x). An outline of such a system is shown in Fig. 1.5a with the azimuth processing performed by an LIV operation whose details we take up in the next section. The recording—e.g., film—which is the output of the terrain imaging system of Fig. 1.4a is no longer an image: it is called, e.g., the *signal film*, and it is the input to the final processing that produces as an output the desired image.

One of the simplest useful analytic models for the synthetic aperture terrain imaging system of Fig. 1.5a is shown in Fig. 1.5b. Such a radar system may include signal dispersion and compression, so this provision is included in the model. The terrain is now convolved in the azimuth dimension not with the antenna pattern b_h of Fig. 1.4b, but rather with $B_h \equiv |B_h| \exp i\varphi$ where φ is the phase modulation due to the Doppler caused by the relative motion. (Of course, the antenna pattern would obviously not be the same function as in Fig. 1.4a if the two systems were designed for the same azimuth resolution, or system azimuth scanning beamwidth.)

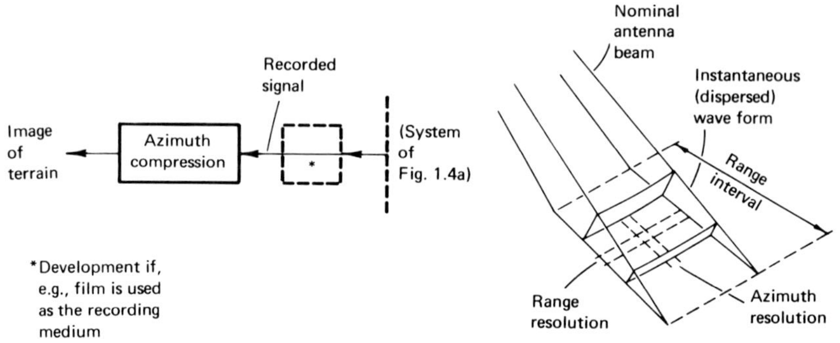

FIG. 1.5a A terrain imaging synthetic aperture radar system (shown employing a dispersed signal waveform, compressed in receiver—see Fig. 1.4a).

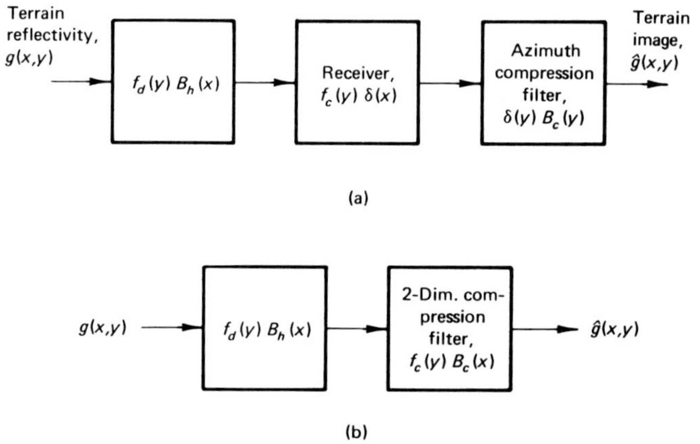

FIG. 1.5b An analytic model for the terrain imaging, synthetic aperture radar system of Fig. 1.5a. The blocks are LIV operations with indicated impulse responses: (a) sequential compression in range (in the receiver in a conventional way) and azimuth; (b) simultaneous two-dimensional compression in range and azimuth.

As is indicated in Fig. 1.5b, there are two different ways to compress the signal. That is, the compression in the range dimension may be done in the radar set's receiver, just as in the conventional range imaging radar system described above, but also, the compression may be performed in *both* dimensions in the radar system's data processor, the final block of Fig. 1.5b. Furthermore, if one employs an *optical* processor, which can operate on a two-dimensional function, the data processor can compress in both *simultaneously*. Still further, the simultaneous, two-dimensional compression is performed effectively *instantaneously* by the optical processing system.

1.4. THE ELEMENTARY OPTICAL PROCESSOR

We have already stated some possible advantages of optical processing of the radar data: the possibility of, essentially instantaneously, simultaneously compressing in both dimensions the entire ensemble of scattered signals from a relatively large area of terrain. These in themselves are very impressive when, as done later, practical designs are contemplated. The use of "instantaneous" optical computation must be qualified in practice to include the time required for the transformation of the received electrical signals into a two-dimensional optical signal; for example, in the commonly used CRT-photographic film recording system, this can be of the order of 1 minute, due essentially entirely to the exposure, development, and (to a small part) the physical transportation of the film.

Another obvious competitor for the data processing role is, of course, the digital computer. But, e.g., the data processing job for a typical orbital synthetic aperture system would require orders of magnitude of time by digital computation greater than by optical system computation. Special purpose digital machines are usable only in very special synthetic aperture systems, and they are likely to be restricted in their range of application for some time to come.

Of course, the digital computer is an accurate, general-purpose machine and is therefore likely to be inferior in some applications for which special types of computing machines can be made. The ability of electronic analog computers, based on the operational amplifier, to solve certain differential equations to suitable accuracy in less time than a digital machine is well known. In the synthetic aperture radar system data processing computation, the coherent optical system is an analog computer: *the optical processor carries out its calculation by physical operations which very closely mimic the actual generation of the radar data from the terrain, whose image we wish to recover.*

To explain the basic idea we consider a single small scattering object whose spatial extent is a few times the projected system resolution. This is logically permissible because we are discussing a linear system and can apply superposition to find the image due to more complicated terrain. The geometry is shown in Fig. 1.6a, adapted from Fig. 1.1. Suppose for the moment that we make the rather unimportant assumption that this object on the terrain is

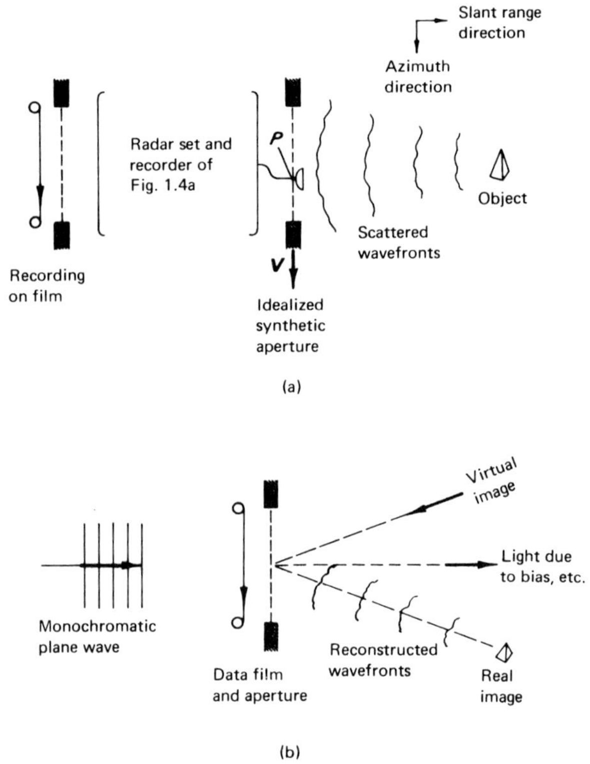

FIG. 1.6 The elementary optical method of processing SAR data in the azimuth direction: (a) generation (construction), sampling, and recording scattered wavefronts. P is instantaneous vehicle position; (b) reconstruction of scattered wavefronts.

illuminated by a plane wave. Then as the vehicle-borne radar set is moved in uniform translation, *the scattered wavefront is sampled periodically over a distance that we term the "synthetic aperture length."* Now to reincorporate immediately the fact that the object is illuminated by a wavefront transmitted from the instantaneous vehicle position, we see by a simple, mental ray tracing that the returned phase is multiplied by two. Further, the synthetic aperture length is determined by the distance over which the *physical* antenna's beam pattern illuminates the object.

Now it is obvious—when stated!—that, if we could somehow recreate a scaled-down replica of this wavefront at optical frequencies, then we could in principle simply use a lens to focus the divergent wavefront and recreate a real image of the object. There is just one difficulty: we must sample, store, and recreate both the amplitude and phase of the scattered wavefront impinging on the synthetic aperture, for otherwise we cannot focus the recreated wave to an image of the object.

From this point of view, the radar receiver and the recorder, along with the attendant vehicle translational motion (Figs. 1.4a and 1.5a), are simply *a mechanism to replicate the wavefront, impinging on the synthetic aperture, onto ordinary photographic film.* The difficulty of preserving amplitude and phase is because the transmission function of film is a function only of the square-magnitude of the impinging light wave. Thus if the impinging wave has a spatial dependence of the form $a(x)\cos\varphi(x)$—ignoring the time dependence for now—then the recorded wave on the film would be, approximately,

$$|a(x)e^{i\varphi(x)}|^2 = a^2(x).$$

Or, if we record on a bias b_o so that $a(x)\cos\varphi(x)$ does not have negative excursions, the recording would be

$$|b_o + a(x)e^{i\varphi(x)}|^2 = |b_o|^2 + a^2(x) + 2a(x)\cos\varphi(x).$$

The first technique is apparently unusable: the phase $\varphi(x)$ of the wavefront has been lost. The second technique—where we have recorded on a bias—does, in fact, contain the desired replica of the wavefront: unfortunately, when we illuminate the exposed film and focus the emergent wavefront, the other two terms will cause a degraded image.

The way around this difficulty is to leave the *electrical* signal on a residual temporal carrier frequency which we will call the "offset frequency." Then the recorded *optical* signal will be on a *spatial* carrier—which we will also call an "offset frequency." Now the recorded wave on the film will be

$$|b_o + a(x)e^{i\omega_o x + i\varphi(x)}|^2 = |b_o|^2 + |a(x)|^2 + 2a(x)\cos[\omega_o x + \varphi(x)].$$

When this replica is illuminated by a plane, monochromatic wave, there will be *three* emerging waves: one due to $[|b_o|^2 + |a(x)|^2]$ which will propagate in the direction of the incident plane wave, as shown in Fig. 1.6b; one (at an angle ω_o/k) due to $a(x)\exp i[\omega_o x + \varphi(x)]$ which is a diverging replica of the original wavefront, and finally one (at an angle $-\omega_o/k$) which converges, or focuses, by itself to an image of the object. *If the offset frequency is properly chosen, then the image will be formed essentially without degrading interference from the other two wavefronts.*†

The idea of an offset frequency has been familiar to electrical, or communication, engineers for a long time. Its recognition within the optical art has caused enormous interest and activity: the wavefront recorded by this technique is called a "*hologram*."

The reader familiar with the hologram technique will notice the close analogy. In the "ordinary hologram" (see Fig. 1.7) all wavelengths are in the

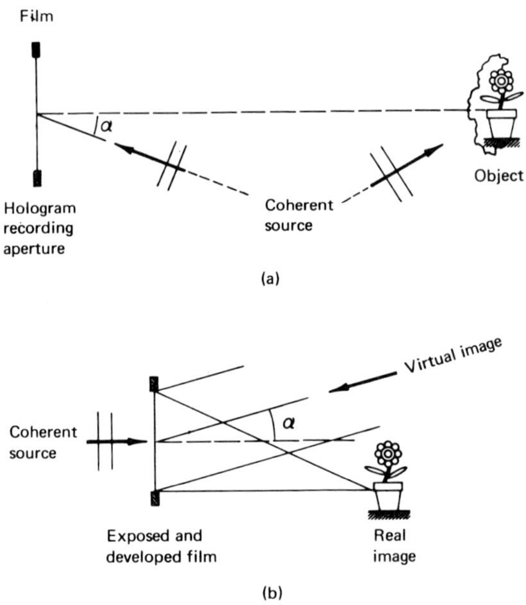

FIG. 1.7 The hologram principle: (a) construction of hologram; (b) reconstruction (imaging) of object.

visible spectrum, and the reference frequency is the illuminating source—collimated into a place wave—and the spatial offset frequency is achieved by *geometrically* arranging to have the reference impinge at a relative angle on the film. It is to be noted that the achievement of the coherent spatial "offset,"

† The association of an angle (α) between the optical axis and the direction of propagation of the wavefront due a spatial amplitude distribution on a carrier/offset frequency ω_o will seem reasonable when we recall that the relation between an aperture distribution and its far field (with the angle α as coordinate) is a Fourier transform—with kernel exp $ik\alpha x$, $k = 2\pi/\lambda$, instead of the usual exp $i\alpha x$. For the Fourier transform would appear as two parts, centered on $\alpha = \pm\omega_o/k$: although we do not observe the wavefronts in the far field, these are their directions of propagation.

or "carrier," frequency is the fundamental invention—not the advent of the LASER—and this recognition was due to E.N. Leith. In fact, the history of the evolution of the essential idea has been briefly sketched in this section and is the very significant discovery of E.N. Leith.

We note that in terms of our previous discussion the recorded signal essentially "compresses" itself: one need only illuminate it suitably, the "compression device" being simple diffraction in *space*. It is difficult to imagine a simpler and more beautiful method of signal processing. As we have said, for many typical SAR's, the modulation due to motion in the azimuth direction is linear FM. When such a signal is recorded (with signal amplitude recorded proportional to transmission of light) onto film, it forms what is known as a "Fresnel zone plate" (in one dimension): it is well known in optical work that when coherent light is passed through such a transmission function it is focussed to a "point" image. If one compares this to electrical compression filters, especially for large duration, large bandwidth signals, it is a truly remarkable simplicity.

1.5. TERRAIN IMAGING RADAR APPLICATIONS

Imaging radar systems, in particular SAR systems, are active sensing devices operating (usually) in the microwave frequency band: this latter fact determines their use vis-a-vis other sensing devices, such as infrared and visible light photography. The commonly known advantage that microwave systems have is greater tolerance to the degradations of the earth's atmosphere, especially those of weather. The less well-known property, that can be either an advantage or disadvantage, of microwave systems is that they "sense" different qualities of the object, both because a different frequency spectrum is used and because a localized illumination is used.

The general advantages are: (e.g., remote) observation location, good resolution at remote distances, large area coverage, rapid data acquisition, and year-round, "all-weather" capability. The general disadvantages are: complex and expensive systems and a requirement for highly trained personnel, including highly trained image evaluators.

We can expect that SAR systems will find applications similar to those already found and projected for visible light photography. Areas of possible application include: geology, soil science, agriculture, hydrology, engineering, disaster analysis, oceanography, and military intelligence.

The general area of remote sensing of the earth appears to be a new area of some promise. There is a voracious appetite in current science and technology for data in the areas of application just listed, and systems (e.g., orbiting) that remotely sense over different segments of the spectral band can be powerful gatherers of such data. It is not clear at the present time just how radar systems such as the SAR system will be combined with other remote sensing systems: experimental work is only just under way to attempt to discover the special virtues of SAR systems in the above areas of application.

1.6. TOPICS TO BE DISCUSSED

Our concern here is with the system engineering of synthetic aperture radar, that is, we wish to discuss the analysis and design of such systems. Such considerations logically begin with the general setting of the system—that is, the function that the system is to perform. To require that the system "make an image of the terrain" is only a modest beginning. A higher level of precision would be achieved if it was concluded that the function was, say, a continuous survey of ocean traffic for use in an economic survey of ocean commerce. These general considerations are more difficult to characterize in their influence on system design, and one often retreats to simpler considerations (such as "resolution," or properties of the system impulse response). Apparently, the attempts to be relatively general are referred to, in the vernacular of the trade, as "system design"; presumably if the attempts become sufficiently schizophrenic they would be termed "engineering design." In fact, there is no logical separation at all, though other limitations may well force a division of labor.

First level of discussion. At the lowest level of discussion we shall ignore the majority of factors that will influence the design of these systems in order to develop, as simply as possible, the relations among and restraints on the principal parameters of the system. In fact, by simple argument we have already noted the importance of the ratio ν/v, where ν is the velocity of light and v the velocity of the vehicle. We adopt, without much ado, the resolution criteria; that is, we examine the error-free impulse response of the system. The concept of resolution, rarely well defined, allows one to proceed rapidly into industrious activity which, fortunately or unfortunately, usually has some relevance to the unknown, or unstated, actual problem. Other criteria, such as least squares estimation, force one to recognize more clearly the "singularity," or unreasonableness, of the error-free model.

The most universally modeled phenomenon whose addition to system models can prevent "singular" behavior is additive random noise. Roughly speaking, this error source is rather "nice," in that, while removing the singular behavior, it still results in about the same processor for a variety of criteria and conditions, and these processors are linear and, frequently, time invariant. Further, analytically it is the most tractable and therefore most fully investigated system error.

For these reasons, we shall have in mind, at least, the presence of such an error in our first level discussion in Chapter II. We should note, however, that in many applications there is plenty of signal power, and therefore the additive noise may *not* be the most important error source. There is therefore no reason, a priori, why the system design should be "warped" to emphasize its presence.

We shall also quickly review the sampling of signals and properties of linear FM waveforms. We will then derive a linear block diagram for the synthetic aperture radar system, which is of paramount importance in the practical design. We shall also note a simple way of estimating a resolution property of the entire system.

Because of the very crucial role played by coherent optical processing in the practical realization of the data processor and in the "natural" position it occupies in that role and, further, because of the present absence of a relevant unifying discussion of this type of processing, we shall devote some time to its exposition in Chapter V.

Second level of discussion. Having rather settled upon a general characterization of the optimum processor, we will now examine its performance in greater detail, with respect to apparently basic limitations and sources of error.

Our discussion above pointed out the similarity of range-dimension compression—commonly referred to as "pulse compression"—and the compression, or processing, of the "physical beam" into the "synthetic beam." It is well known that pulse compression is rather sensitive to multiplicative errors, most commonly phase errors. Intuitively, this is because the phase errors directly distort the phase modulation which provides the "fine structure," upon which the compression operation depends. Therefore, we shall devote some time to the discussion of the analysis, both analytical and by "random experiment," of this very important effect in Chapter VI.

Another error source is departure from uniform translation motions that are characterized by changes in the magnitude—but not the direction—of either the vehicle, the recording film, or the recording sweep. These "motion errors" bear some resemblance to phase errors but are a separate class that requires attention. We shall derive some important tolerable limits on these errors, design requirements, and even a method of testing for these errors (Chapter VI).

It is well known that the received waveforms scattered by terrain have a very large range of possible amplitude levels relative to the range of levels that can be accommodated linearly by practical receiving, recording, and processing equipment. The synthetic aperture radar has the curious possibility of reducing this so-called "dynamic range" by its very method of generating the data in a dispersed form. But it is still of considerable interest to discuss the effect of nonlinearities on the radar performance. Fortunately, the more well-investigated "no-memory" model of nonlinearities suffices to explain most of the nonlinear phenomena—such as weak signal suppression and generation of false images—that occur in these systems (Chapter VII).

One of the physically most obvious sources of error is the "nonideal" shape of the physical antenna pattern in its far field. This pattern affects the relative size of "ambiguous," or false, images on a large scale and, in the azimuth dimension, the local, or sidelobe, behavior of the system impulse response. The signal modulation determines the local, or sidelobe, behavior of the system impulse response in the range dimension. This latter (range dimension sidelobes) behavior is essentially independent of the first for the normal geometries and for periodic pulse modulation. We shall see that the "ambiguity function" of the SAR is not the same as the "classical" ambiguity function for a range-Doppler radar—indeed there is no reason why it should be—and that, unlike the "classical" radar, the SAR possesses an inherent suppression of range ambiguties. The optimum design of the antenna pattern and signal modulation, along with the pulse repetition frequency which necessarily is involved, is therefore discussed at some length in Chapter VIII.

Third level of discussion. Finally, we shall discuss what influence, if any, the effects of the error sources investigated at the second level should have on the design of these systems. We shall find these questions much more difficult to answer and requiring more careful formulation; the answers represent, roughly speaking, more complicated and less flexible data processing schemes.

One of the more obvious problems of this type is, to state it one way, the rigid constancy imposed on the range interval to azimuth resolution ratio by the "simple" system. The formulation of the problem is at least a little subtle because there are obvious ways to subvert this limitation if one takes a sufficiently detached view of it: e.g., one could place two independent radar sets on the vehicle and, with some care, eliminate redundant parts. We shall discuss in detail why some obvious "solutions" fail. We shall also prove that it is not possible to choose a signal modulation that can *avoid* this restraint, when the processor is that one optimum for additive, white noise. We will observe that, however, it is possible to *suppress* the range ambiguities by choice of signal modulation to a level presumably useful for many applications. In the course of this discussion, numerous properties of the ambiguity function for this radar system will be established (Chapter IX).

We next discuss the design of the optimum processor when at the outset we include phase errors in the signal channel model. We shall obtain results under the resolution, maximum SNR, and least-squares criterion—assuming a linear processors—and discuss them at some length. We shall also consider maximum likelihood processors; they turn out to be nonlinear, and their analysis is correspondingly difficult (Chapter X).

COMMENTS AND REFERENCES

Conventional radar. There is an enormous literature on radar systems. A "classical" compendium is the famous 28 volume M.I.T. "Rad-Lab" series, several of which remain admirable discussions; a good introduction to radar is Ridenour (1947). The system discussion of radar is usually oriented toward the detection and parameter estimation of one or a few scattering objects; a recent exposition of this type of radar is given in Berkowitz (1965) and a discussion of the linear FM signal and its system was discussed at length by Klauder *et al.* (1960).

Synthetic aperture radar. The first reporting of successful imagery obtained by the synthetic aperture technique was by Cutrona *et al.* (1961). A set of papers discussing the early development and giving rudimentary analyses was Sherwin *et al.* (1962), Cutrona and Hall (1962), McCord (1962), and Heimiller (1962). The paper of Sherwin *et al.* gives credit to Carl Wiley, then of Goodyear Aircraft Corporation, for advancing and analyzing the idea that improved angular resolution can be possible because of the Doppler frequency structure. More recently, tutorial discussions in some detail of the technique and coherent optical processing has been given by Cutrona *et al.* (1966), and Brown (1967).

Holography. The elegant ideas of Gabor and Leith are not essential but very instructive in understanding and generalizing the synthetic aperture principle; see Leith and Upatnieks (1962).

Optical processing. The techniques and literature on data processing by coherent optical methods is undergoing explosive expansion. A relevant review of the basic ideas is given in Cutrona *et al.* (1960).

Applications. The use of the SAR technique will invariably be similar to those of aerialphotography; a review of applications of the latter technique appears in Colwell (1966).

II

Synthetic Aperture Principle and Radar : Simplified Discussion

2.1. INTRODUCTION

In this chapter we begin the discussion of the synthetic aperture radar system in the simplest manner possible. We shall ignore all subtleties that we can—which is nearly all of them!—in order to establish the principle parameters of the system and the relations among them.

In Sec. 2.2 we shall examine the scattered signal, or "azimuth history" as we shall sometimes refer to it, for a single scatterer of very small spatial extent, assuming the most typical geometry. As we do not worry about range measurement in this section we may assume monochromatic ("CW") radiation. We shall point out—delaying the details until Sec. 2.5—that, again for the typical geometries, the spectrum of the "azimuth history" can be determined in a simple way—by the so-called method of stationary phase—that is extremely convenient in analysis and very useful in providing an intuitive understanding of this type of signal and, importantly, subsequent operations of various types upon it. In order to characterize the "compression" of the "azimuth modulation"—as we prefer to call it, instead of "azimuth history"—we will have to be a little careful. We shall assume the presence of additive white noise in order to avoid anamolous, or "singular" answers. In effect, in this section we are analyzing the impulse response of the system in the azimuth direction.

In Sec. 2.3 we shall make a terrain imaging radar based on the synthetic aperture principle in the apparently simplest possible way: by using a periodic pulse modulation. With this type of radiated waveform the Doppler modulation is sampled: well-known requirements on sampling rate lead to seemingly fundamental restraints on the parameters, or operation, of this type of radar. While stressing the qualitative aspects of the discussion we will derive the basic relations among, and restraints on, the parameters bandwidth, resolution, sampling frequency ("PRF"), antenna size, and unambiguous range interval.

We will have observed in Sec. 2.2 that the only way the azimuth position of the scatter is known is via its azimuth modulation due to the Doppler effect. This leads us to develop the idea of the "Doppler space" about the radar-carrying vehicle. One of the major problems of synthetic aperture radar systems is analyzing and controlling the relation between these two coordinates. The concept of a Doppler space will allow considerable insight into the complications due to sampling, vehicle motion, terrain motion—e.g., planetary rotation, eccentricity of vehicle orbit, and so forth. By also considering the range measurement, or imaging, we can easily describe the gross nature of the mapping operation (in the mathematical sense) from object to image accomplished by the SAR system.

2.2. THE SYNTHETIC APERTURE PRINCIPLE

The physical phenomenon employed by the synthetic aperture radar to achieve azimuth resolution, or fidelity, is the Doppler effect, as we have already noted above. As a source (or scatterer) passes by a receiver (or radiator-receiver), the frequency of the received waveforms decreases. Qualitatively, one may say, in the vernacular of the sonar trade, that, as the source/scatterer passes abeam, the situation changes from "up Doppler" to "down Doppler." Measuring the time of this transition, therefore, gives an angular coordinate of the source/scatterer at that time. The additional measurement of the range at that time gives the polar coordinates of the source/scatterer, with the radar-set-carrying vehicle† at the origin of the coordinate system.

The "history," or Doppler modulation, is determined by the geometry of source/scatterer and receiver configuration. The most typical geometry for the SAR has been shown in Fig. 1.1. When we are considering the return of a single scatterer, it is convenient to resketch the geometry in the plane containing the vehicle and the scatterer and containing the vehicle velocity vector. This is shown in Fig. 2.1. We shall refer to this plane as the "slant range plane."

The received signal. Assume the radar set radiates a monochromatic (CW) wave whose time behavior is therefore simply given by the function $\cos \omega_0 t$. A small portion (spatially) of the radiated wavefront will be scattered by the small object we are considering; the scattered wavefront is then incident

† From here on, the unmodified use of vehicles will always mean that mechanism that is transporting the radar set.

upon the receiving aperture, producing in the radar set an electrical signal of the form $\mathcal{C} \cos \{\omega_c[t - \tau(t)]\}$ where $\mathcal{C}$ is an amplitude scale,† and $\tau(t)$ is the delay due to the time required for propagation. (We ignore all other delays since they generally are constant, independent of time, and will be eventually accounted for by assuming the received signal has an unknown constant phase.)

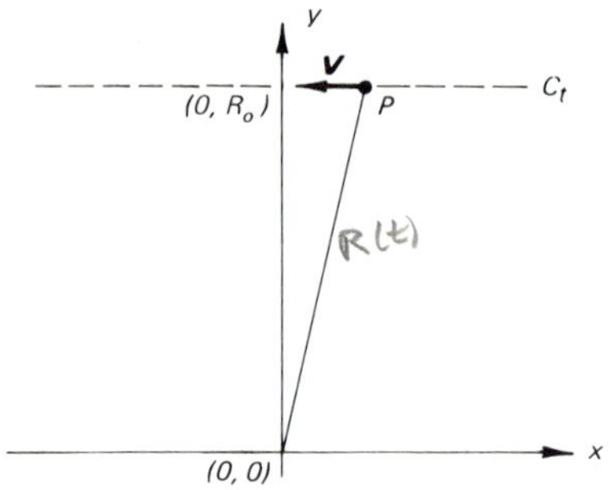

FIG. 2.1 The typical geometry of the SAR in the slant range plane. P is the scattering object's instantaneous relative position; **v** its relative velocity vector, and C_t its relative trajectory.

The delay $\tau(t)$ is equal to $2R(t)/v$ where $R^2(t) = R_o^2 + (vt)^2$ as is obvious from Fig. 2.1. In the most typical SAR configuration the antenna azimuth pattern is such that the signal is received primarily over a rather small angle, centered abeam. Therefore, to a generally good approximation $R(t) \approx R_o + v^2t^2/2R_o$ [by the binomial expansion of $(a + b)^{1/2}$]. Thus the phase modulation due to the Doppler effect caused by the relative motion is

$$\varphi(t) \approx 2k_\lambda R_o + \frac{k_\lambda v^2 t^2}{R_o} \tag{2.1}$$

where the "wavenumber" $k_\lambda \equiv 2\pi/\lambda$, where $\lambda \equiv 2\pi v/\omega_o$ is the wavelength of the monochromatic radiated wave. The received waveform is then approximately

$$s(t) = \mathcal{C}\,|A(t)| \cos\left(\omega_c t - \frac{4\pi R_o}{\lambda} - \frac{4\pi v^2}{\lambda R_o} \frac{1}{2} t^2\right), \qquad |t| < \tfrac{1}{2}T. \tag{2.2}$$

T is the length of time over which we receive the azimuth history; as we have assumed reception over a small angle, we must have $X/R \ll 1$ rad, where $X \equiv vT$ is the spatial interval over which we collect the scattered wavefront.‡ The amplitude modulation $|A(t)|$ is due to the physical antenna pattern in the far field. The constant $k \equiv 4\pi v^2/\lambda R_o$ we will call the "linear FM rate."

† We shall employ the symbol "$\mathcal{C}$" in the remainder to indicate the presence of an amplitude scale (often complex) whose exact nature or value is not of immediate interest.

‡ We will discuss the interesting situation when the Doppler modulation is received over nonsmall angles in Sec. 2.5.

Therefore, as was asserted above, in the most common SAR configuration. the phase modulation due to the Doppler effect is approximately linear FM. As is generally true, in order to describe more easily the effect of linear, time-invariant filters on this signal, its spectrum—i.e., Fourier transform—is of interest. The exact calculation of the spectrum involves Fresnel integrals, which are numerically tedious to work with. Fortunately, for the usual SAR configuration, the method of stationary phase provides a useful approximation. This method is so essential to an intuitive understanding of the SAR that it is worth discussing in some detail in the present context; however, we shall delay such discussion until Sec. 2.5 in order not to obscure the derivation of the basic SAR parameters here. Using this approximate method we can quickly write down the spectrum of the azimuth modulation:

$$\tilde{S}(\omega) \approx \mathcal{C}\left|A_T\left(\frac{\omega+\omega_o}{\Omega/T}\right)\right| \exp\left[\frac{-i(\omega+\omega_o)^2}{2k}\right] + \mathcal{C}\left|A_T\left(\frac{\omega-\omega_o}{\Omega/T}\right)\right| \exp\left[\frac{i(\omega-\omega_o)^2}{2k}\right], \quad (2.3)$$

where $|A_T(t)|$ is $|A(t)|$, except it is zero for $|t| > \frac{1}{2}T$. The signal bandwidth is approximately the difference Ω of the instantaneous frequencies at $t = \pm\frac{1}{2}T$:

$$\Omega \equiv \frac{4\pi v^2 T}{\lambda R} \ \frac{\text{rad}}{\text{sec}}. \quad (2.4)$$

The signal $s(t)$ is sketched, in distorted detail, in Fig. 2.2a. A typical SAR waveform would have hundreds or thousands of zero crossings due to its modulation alone and, of course, an arbitrarily large number when modulated on a carrier of arbitrary frequency ω_o. The magnitude, $|\tilde{S}(\omega)|$, of the signal spectrum is shown in Fig. 2.2b, using the method of stationary phase; it is principally determined by the antenna pattern in azimuth.

Signal compression. We now wish to process the received data in order to estimate the time at which the scatterer passed, say, abeam. For then, if we know the vehicle's position with respect to a terrain coordinate system at that time, we can estimate one of the scatterer's terrain coordinates. Actually, in many applications we do not care about more than a rough relation to the terrain coordinates but rather desire the image to have good positional accuracy "internally." Or to say the same thing another way, we wish to record the signal on a (two-dimensional) format—say photographic film—and estimate position as precisely as possible relative to a rather arbitrary coordinate system on the format.

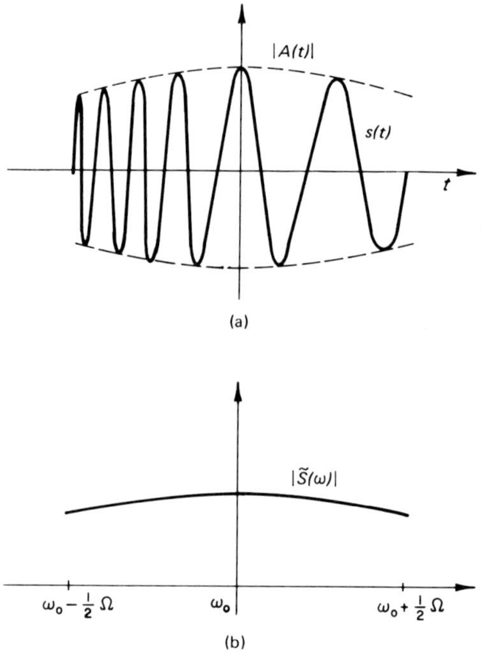

FIG. 2.2 A linear FM signal (a) and its spectral magnitude (b) in the large TBP case.

But if the received data is comprised solely of the received signal [Eq. (2.1)], this time—hence position—can be measured *precisely*. This conclusion is repugnant for several reasons. First of all, the accumulated experience in physics concerning measurement of physical quantities indicates that precise measurement is never possible, and models that permit it are deficient in some important respect. The most common amelioration of this difficulty in nonatomic scale problems is to assume the presence of random effects which are additive; an appeal may or may not be made to a "central limit theorem."

Secondly, in the SAR system we know very well that we must be concerned with processing a great number of signals. For example, in the present context we are interested in estimating the positions of a great number of small scatterers. Even if their positions could be measured precisely, we would need to implement a method of accomplishing it. From our prior discussion we feel intuitively that "compressing" all these signals, so that they cause a response, or image, mainly concentrated near where "they are," is certainly a worthwhile initial processing, perhaps leaving a final interpretation to a vaguely guided human observer.

We shall defer more precise arguments until Chapter IV and state that there we show that, on the basis of several viewpoints, a good processing scheme is to employ the so-called "matched filter" as at least part of the processor. For example, if we really are concerned with estimating the time at which a small scatterer passes abeam, we should pass its signal through this "matched filter" and take the time of occurrence of the maximum of the envelope of the output of the filter as the estimate of the abeam time.

The "matched filter" is defined to have a transfer function—which, of course, is the Fourier transform of the impulse response—which is the complex conjugate of the signal spectrum. Thus the transfer function is—from Eq. (2.3)—

$$\bar{\tilde{S}}(\omega) = \bar{\mathcal{C}} \left| A_T\left(\frac{\omega + \omega_o}{\Omega/T}\right)\right| \exp\left[\frac{+i(\omega+\omega_o)^2}{2k}\right] + \bar{\mathcal{C}} \left| A_T\left(\frac{\omega - \omega_o}{\Omega/T}\right)\right| \exp\left[\frac{-i(\omega-\omega_o)^2}{2k}\right]. \quad (2.5)$$

We note in passing a fact that complicates somewhat the design of the processor when the real problem of imaging a certain range interval is taken up: since k, upon which $\tilde{S}$ depends, is equal by definition to $4\pi v^2/\lambda R_o$, the processor must realize a matched filter that is a function of slant range R_o.

The signal output of the matched filter has, of course, the spectrum $|\tilde{S}(\omega)|^2 \equiv \tilde{S}_o(\omega)$; the output signal as a function of time is therefore given as the inverse Fourier transform of $\tilde{S}_o(\omega)$. To keep the present development simple, suppose that we assume a rather "ideal"—and impossible!—rectangular far-field pattern in azimuth for the antenna so that $|A_T(t)| \equiv 1$, $|t| < \frac{1}{2}T$.

Then, as long as $\omega_o > \frac{1}{2}\Omega$ we have, within an amplitude scale,

$$\tilde{S}_o(\omega) = \begin{cases} 1, & -\omega_o - \frac{1}{2}\Omega < \omega < -\omega_o + \frac{1}{2}\Omega, \\ 1, & \omega_o - \frac{1}{2}\Omega < \omega < \omega_o + \frac{1}{2}\Omega, \\ 0, & \omega \ \text{otherwise}. \end{cases} \quad (2.6a)$$

$\Omega = \frac{4\pi v^2 T}{\lambda R}$

The calculation of the inverse transform is rather simple:

$$S_o(t) = \mathcal{C}\,\frac{\sin \frac{1}{2}\Omega t}{\pi} \cdot \cos \omega_o t. \quad (2.6b)$$

So the desired output, the squared envelope, is

$$S_{oe}^2(t) = \mathcal{C}\left(\frac{\sin \frac{1}{2}\Omega t}{\pi t}\right)^2. \quad (2.6c)$$

This is the response of the SAR, under the above assumptions and approximations, to a small scattering object; it is plotted in Fig. 2.3.

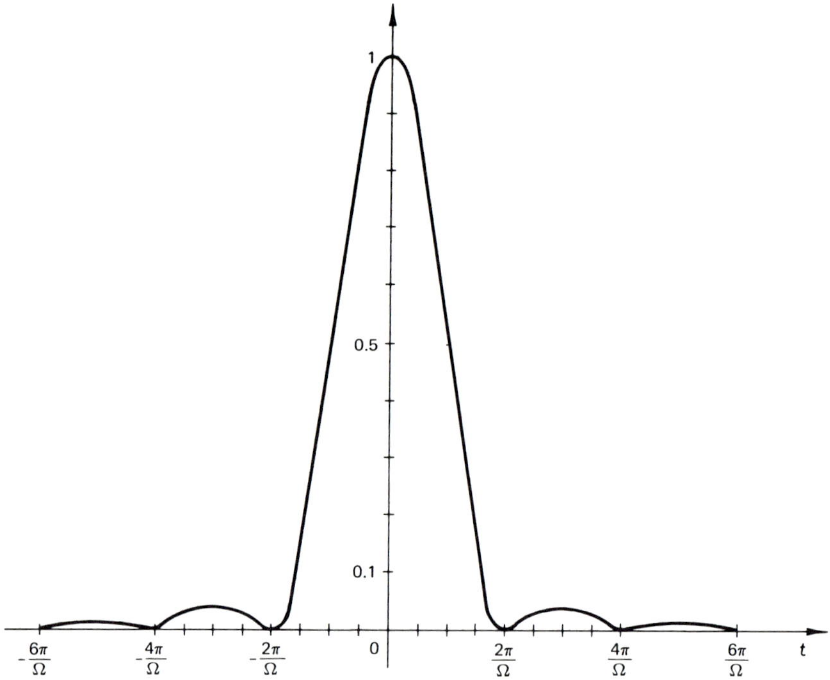

FIG. 2.3 The functional form of the approximate output image, azimuth cross section; of a small scatterer, under idealized assumptions.

A useful and often used measure of how well the system can distinguish between closely separated small scattering objects—by producing an image with two distinct "peaks"—is given by the "spread," suitably defined, of the impulse response. A commonly used measured of "spread" is the width, e.g., in units of time or space, of the impulse response between the points where the response has (first) decreased to $1/\sqrt{2}$—"half power"—of its maximum value. Such a measure of spread is commonly referred to as the "resolution" of the system. From Fig. 2.3 we have the "3 db resolution," if $\Omega_f \equiv \Omega/2\pi$,

$$\rho_a = \frac{0.88}{\Omega_f} = 0.88\,\frac{\lambda R}{2v^2 T} \quad \text{sec}$$

$$= 0.88\,\frac{\lambda R}{2vT} \quad \text{m.} \qquad (2.7)$$

We observe that the resolution is inversely proportional to signal bandwidth: this is a general "rule of thumb" in fact. Because of the linear FM modulation the bandwidth is proportional to signal duration T [e.g., Eq. (2.4)], and, therefore, the resolution is inversely proportional to T. This is a specific proof of the claim that it is not necessary to have a narrow physical beam to obtain a system resolution attributable to such a beam.

But it is obvious that the ability of a system to distinguish closely spaced objects—that is, its "resolution ability" or "resolving power"—cannot be adequately described by a single number. Consider the response shown in Fig. 2.4: the so-called "sidelobes" on the impulse response are so great

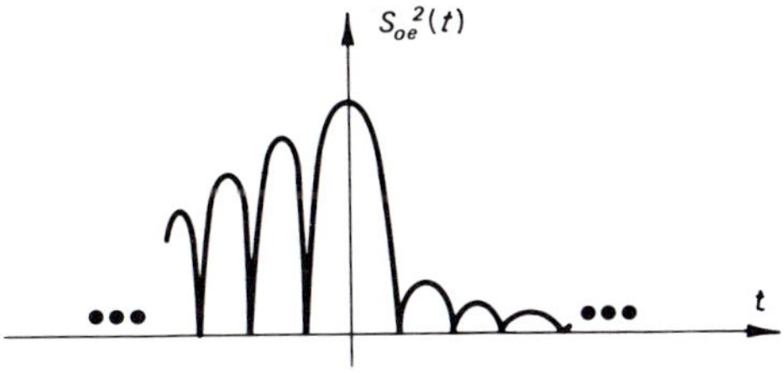

FIG. 2.4 A system response whose resolution property is not well measured by a measure of duration of the principal lobe.

that they often cause hopeless confusion in the response of closely spaced scatterers. In fact, just such a response can be caused by the presence of phase errors in an otherwise well-designed system (Sec. 6.2.1); that is, there can be ambiguity between the detail of an image and phase error effects. It is therefore obvious that a measure of "sidelobe level," defined, say, as the relative height of the second greatest maximum, is equally important with "spread." It is well known, and calculable directly from Fig. 2.3, that the sidelobe level of the $(\sin x)/x$ type response is 13.4 db.

One can pursue the problem of a meaningful definition of resolution still further, of course, but we shall defer such discussion until Chapter IV.

As we have already noted above, the length of the time T that the scattered signal is received is determined effectively by the azimuth pattern of the physical antenna. Suppose that the (physical) antenna has aperture length D_h (in the azimuth dimension) and uniform illumination and is boresighted exactly abeam. The far-field pattern will be of $(\sin x)/x$ form with a 3 db beamwidth of $0.88\,\lambda/D$: for simplicity we approximate this pattern to be unity within this angular interval and zero outside it. Then the results immediately above, Eq. (2.6c) and Fig. 2.3, apply with now

$$T \equiv \frac{(0.88\lambda/D_h)\,R}{v} \quad \text{sec} \tag{2.8}$$

and so the azimuth 3 db resolution

$$\boxed{\rho_a = \frac{D_h}{2} \quad \text{m.}} \tag{2.7a}$$

In words, the azimuth resolution for the typical SAR configuration is approximately half the azimuth dimension of the aperture of the physical antenna. It is important to note that, as we claimed above, *the azimuth resolution is independent of all other parameters, including range and wavelength.*†

We note also that with T determined in this manner, the bandwidth of the azimuth modulation becomes

$$\boxed{\Omega_f = 0.88 \frac{2v}{D_h} \quad \text{Hz.}} \tag{2.4a}$$

2.3. SYNTHETIC APERTURE RADAR

A synthetic aperture radar combines the azimuth measurement method, based on the synthetic aperture principle we have just discussed, with some method of determining range. The simplest scheme is to pulse modulate the monochromatic wave radiated by the radar set, in much the same manner as "conventional" radar systems. This implies that the azimuth modulation, or Doppler history, will now be *sampled* instead of being received continuously: in order to recover the (continuous) azimuth modulation we must sample at a sufficiently high rate, as is well known.

We will show (Sec. 3.4) that when the radiated modulated monochromatic waveform is a periodic sequence of (identical) pulses, then the range and azimuth measurements are approximately independent, assuming the typical SAR configuration and matched filter processing. Therefore, the design of the pulse modulation to achieve resolution, or fidelity, in the range dimension does not differ from the design of pulses for range measurement only. This latter problem has, of course, been extensively investigated since the inception of radar. In particular, it is well known that the resolution and accuracy of the range measurement, or the fidelity of the range image, is proportional, roughly, to the bandwidth and energy of the pulse modulation. The design problem is in part a compromise between resolution and sidelobe level. One can expect peak instantaneous radiated power limitations in some applications that will force the use of dispersed pulse modulations—e.g., linear FM, or "chirp" modulation—in order to obtain the required bandwidth and energy.

If, in fact, linear FM is used as the signal modulation, we note that there is a marked similarity between the azimuth and signal—or "range"—modulations and their required matched filters. If a periodic pulse modulation is

† We shall consider more carefully when the assumptions hold, and the interesting behavior when they do not, in Sec. 2.5.

used, these signal modulations and their filtering may be analyzed independently, giving rise to the very useful simplification of an *azimuth* (*signal*) *channel* and a *range* (*signal*) *channel*. We shall develop this idea in Sec. 3.5.

This method of measuring range is a mixed blessing; although simple (because familiar), it introduces the also familiar range ambiguity problem and some complicated effects on azimuth measurement performance because of the sampling operation. The range ambiguity problem is partially alleviated in this type of radar in its usual geometrical configuration for two reasons. First, since the vehicle is elevated above the ground, one can design a vertical antenna aperture illumination to illuminate preferentially the unambiguous range interval; since the aperture width is necessarily finite, it is not possible to restrict the far-field antenna pattern entirely to the (finite) range interval. Second, the azimuth history of a range ambiguous scatterer differs from that of the unambiguous scatterer [as we noted after Eq. (2.5)], so a properly designed processor will not respond as well to the ambiguous target; i.e., it suppresses the range ambiguity in the output image.

In order to have a usefully large unambiguous range interval, the sampling rate must not be required to be too large; in practice, this means that there will necessarily be some azimuth signal energy aliased, a well-known phenomenon. This is the manner in which azimuth ambiguous targets manifest themselves in this radar. It also means that the requirement for good azimuth measurement (e.g., resolution) conflicts with the requirement for large range coverage: this is a fundamental system design problem of synthetic aperture, side-looking radars.

It is useful at this point to simplify the real design problem by assuming that the far-field antenna pattern is rectangular in azimuth and elevation, of width equal to about the 3 db width of a uniformly illuminated aperture of the same size. Then it will be possible to derive many important system parameter relations in a simple way. We arbitrarily start with a horizontal, or azimuth, aperture width D_h; then the bandwidth of the azimuth signal due to a scatterer at slant range R_s is, from Eq. (2.4a)—ignoring the factor 0.88—

$$\boxed{\Omega = \frac{2v}{D_h} \quad \text{Hz.}} \tag{2.4b}$$

As is quite well known, in order to sample properly a modulation with a bandwidth of Ω, the sampling rate, f_s, must be at least Ω. As we will discuss at length (Sec. 3.3), there are two natural processing modes for this type of signal: in the "range offset" mode, $f_s = \Omega_f$, and in the "azimuth offset" mode, $f_s = 2\Omega_f$. The names are derived from the fact that in the first mode the signal is processed on a residual carrier frequency equal to about half the signal, or range, modulation bandwidth, while in the second mode the signal is processed on a residual carrier frequency equal to about half the Doppler, or azimuth modulation bandwidth. Thus

$$f_s = \begin{cases} \dfrac{2v}{D_h}, & \text{range offset mode,} \\[2ex] \dfrac{4v}{D_h}, & \text{azimuth offset mode.} \end{cases} \tag{2.9}$$

The sampling frequency, often called the pulse repetition frequency and denoted PRF, limits the range interval over which only a single pulse can be received. In "classical" radars this interval always had the radar set at its near extreme, and thus the far extreme was its "maximum unambiguous range," in the sense that, if objects beyond this range scattered a perceptible return, they would give an image, or return, at an apparent range less than its "maximum unambiguous range." We shall call the range interval determined by one-half the distance light can travel in the time between pulses the *unambiguous* (*slant*) *range interval*, ΔR. It should be noted that there is no necessity for the radar set to be at the near edge of this interval in an SAR system: the recorder CRT, e.g., need only be synchronized to begin its sweep at the time the scattered signal from the near edge of ΔR is received.† We have $\Delta R = \mathrm{v}/2f_s$ so that

$$\Delta R = \begin{cases} \dfrac{\mathrm{v}D_h}{4v}, & \text{range offset mode,} \\[2ex] \dfrac{\mathrm{v}D_h}{8v}, & \text{azimuth offset mode.} \end{cases} \tag{2.10}$$

The geometrical ideas here, though simple, are clarified by reference to Fig. 1.1.

For any given depression angle γ_o and vehicle altitude h, the unambiguous slant range interval ΔR determines an *unambiguous ground range interval* ΔR_g that is to be illuminated preferentially by the physical antenna. The angle β_v subtended at the vehicle by ΔR_g is given implicitly by the equation

$$(2h \cos \gamma_o) \sin \tfrac{1}{2}\beta_v - \tfrac{1}{2}\,\Delta R \cos \beta_v = -\tfrac{1}{2}\,\Delta R \cos 2\gamma_o,$$

† If the sweep begins at some other time, but recurs with the same frequency f_s, the image of the terrain will be divided into two range segments which appear inverted in range but are otherwise normal images.

where γ_o is the depression angle between local horizontal and the antenna boresight. (We have assumed the terrain to be a flat plane normal to the local vertical through the vehicle.) When, as is often true, $\beta_v \ll 1$ rad we can solve this transcendental equation and find

$$\beta_v \approx \frac{\Delta R \tan \gamma_o}{R_m}$$

where R_m is, for example, the mean distance, or slant range, to the illuminated terrain.

If we take λ/D as the antenna pattern beamwidth associated with an aperture width D, then the aperture width required, at wavelength λ, to confine its pattern to the angle β_v approximately will be called the "vertical† aperture width, D_v," and is given by

$$D_v = \begin{cases} \dfrac{8vR_m}{D_h f_o \tan \gamma_o}, & \text{range offset mode,} \\[2ex] \dfrac{4vR_m}{D_h f_o \tan \gamma_o}, & \text{azimuth offset mode,} \end{cases} \tag{2.11}$$

where $f_o \equiv \omega_o/2\pi = v/\lambda$. In order to avoid range ambiguities, we therefore need D_v at least as large as the right sides of Eq. (2.11).

We can put these results in interesting forms. For example, for unambiguous operation for a synthetic aperture sidelooking radar we must have

$$D_v D_h \geq \begin{cases} \dfrac{4vR_m}{f_o \tan \gamma_o}, & \text{range offset operation,} \\[2ex] \dfrac{8vR_m}{f_o \tan \gamma_o}, & \text{azimuth offset operation,} \end{cases} \tag{2.12}$$

a lower bound on the required antenna aperture area.

We may combine the relations for azimuth resolution ρ_a and unambiguous slant range interval ΔR to get

$$\frac{\Delta R}{\rho_a} = \begin{cases} \dfrac{v}{2v}, & \text{range offset,} \\[2ex] \dfrac{v}{4v}, & \text{azimuth offset.} \end{cases} \tag{2.13}$$

† "Vertical" is the direction normal to the plane of the antenna boresight and vehicle velocity vector; it coincides with "local vertical" only if $\gamma = 0$.

That is, *the ratio of the unambiguous slant range interval and the azimuth resolution is a constant fixed by vehicle velocity*. For example, as the desired azimuth resolution becomes smaller, the unambiguous slant range interval becomes smaller. This is the fundamental system parameter restraint for synthetic aperture radar. For this reason it is good to review the assumptions that led to it:

1. The transmitted signal was a coherent sequence of identical pulses.
2. The antenna azimuth aperture length was fixed at twice the azimuth resolution distance.
3. Less fundamentally, an idealization was made concerning the far-field azimuth antenna pattern.

We shall consider the possibility of avoiding this restraint in Chapter IX. Of course, there are some obvious ways to subvert the restraint. We could map a slant range interval $2\,\Delta R$ simply by making two systems or, for that matter, use the same system on the separate mapping runs: the first solution is uneconomical (in several senses), the second often operationally undesirable, and neither is clever (presumably a virtue).

2.4. THE SAR DOPPLER SPACE

We have claimed, and will show in Sec. 2.5, that the spectrum of the azimuth modulation can be easily and well approximated by the method of stationary phase for the typical SAR configuration. The very important consequence of this fact is that for signals of large time (extent)-bandwidth product, denoted TBP, there is a useful relation between the instantaneous frequency of the modulation and its spectrum. For example, the bandwidth of the signal modulation is approximated by the difference of the extremes attained by instantaneous frequency, assuming that the signal amplitude modulation gives it a fairly well-defined time extent—as is true here, the amplitude modulation being determined by the physical antenna pattern. It is also true that the *maximum output response, i.e., approximate image position, is determined by the time at which the signal's instantaneous frequency passes (or would pass if continued in time) through the center frequency of the matched filter*. Because of this, a system of coordinates of constant instantaneous frequency, or constant Doppler, can be a source of quick information about the ensemble of signals that are selected by the physical antenna pattern.

A vehicle moving in uniform translational motion relative to the terrain can be considered to establish a family of surfaces of constant instantaneous Doppler frequency. Since the Doppler depends solely upon magnitude of the relative radial velocity vector, which in turn depends solely upon the angular position relative to the vehicle's relative velocity vector, the surfaces are cones: their apexes lie at the position of the vehicle, and their axes are coincident with the vehicle's relative velocity vector. *This is the natural angular coordinate system of the SAR and is the sole way the angular coordinate of a scatterer is established.*

The measurement of range provides a family of surfaces of constant distance from the vehicle—i.e., a family of spheres concentric about the vehicle. (The intersections of these two "orthogonal" families of surfaces are circles, but this does not interest us here.)

If we consider the terrain to be a planar surface of varying reflectivity, then the intersection of this plane with the family of cones of constant Doppler gives *curves of constant Doppler*: they are by definition "conic sections, or curves" and in situations of interest here will be rectangular hyperbolas. The intersection of the terrain plane with the surfaces of constant range will, of course, be circles of constant range.

On the basis of this idealized terrain model we can get a good idea of how the SAR system forms its image—i.e., how it transforms the (assumed for now) two-dimensional terrain surface into a two-dimensional image. The mapping is not one-to-one, of course: because of the nonzero spread of the system impulse response, the mapping is "one-to-many"; that is, the reflectivity of one incremental area is "blurred" into many other incremental areas. And, as we shall discuss further below, the mapping is also "many-to-one"; that is, many incremental areas can be superimposed in an incremental area of the image due to ambiguities in range and azimuth, certain three-dimensional terrain, and object motion—the latter obviously interferes with the assumed and utilized relation between instantaneous Doppler frequency and azimuth position. Nevertheless, by ignoring these effects, we can get a useful insight into the gross nature of this mapping.

Very low altitude vehicle. As a simple example to illustrate these ideas, we consider the situation where the terrain, vehicle, and vehicle velocity vector are coplanar. This is not typical, of course, of an SAR geometry but approximates the situation when the range interval to be imaged is at a slant range which is large relative to the altitude of the vehicle above the (mean) terrain plane.

Since now the terrain plane contains the axis of the conical surfaces of constant Doppler, the curves of intersection are straight lines through the vehicle position, a degenerate set of rectangular hyperbolas. From Eq. (2.1) we have, setting $y = R_o$,

$$\frac{d\varphi}{dt} \equiv \omega_d(x, y) = \frac{4\pi v x}{\lambda y}, \qquad x \equiv vt.$$

We keep in mind the assumption of small angles in azimuth. The equation $d\varphi/dt = \omega_*$, $x = vt$, yields the equation of these straight lines:

$$x = \frac{\lambda\omega_*}{4\pi v} \cdot y.$$

The intersections of the family of spheres of constant range with the terrain plane are circles concentric to the vehicle position:

$$x^2 + y^2 = R^2.$$

This family of curves, or coordinate system, is shown in Fig. 2.5. Any particular antenna pattern (in azimuth), in any boresight direction, can be superimposed upon this figure: the Doppler histories, bandwidths, center

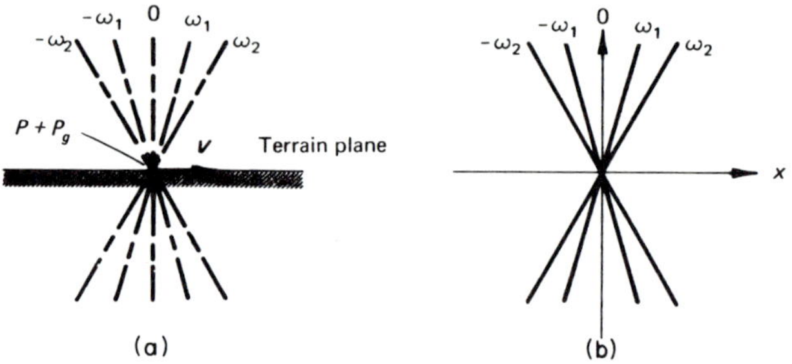

FIG. 2.5 Constant Doppler curves for very low altitude vehicle: (a) curves of constant Doppler: intersection of cones of contant Doppler and vertical plane (defined to contain vehicle velocity vector and local vertical); (b) curves of constant Doppler: intersection of cones of constant Doppler and terrain plane.

frequencies, etc., can be seen by inspection. We observe, for reasons that will become clearer, that the zero Doppler curve is a straight line orthogonal to the vehicle velocity vector, rather obviously in this simple case. Thus if this direction coincides with the boresight direction of the antenna pattern in azimuth, then the whole ensemble of selected signals will have the same center frequency: this is important in the design of the processor. We note that in this simple geometry no matter what the azimuth boresight of the antenna pattern, the Doppler histories will all have the same bandwidth and center frequency: this is not generally true. In fact, the bandwidth, but not center frequency, is independent of the antenna pattern azimuth boresight. Finally, we note that the time duration of the Doppler histories is linearly proportional to range: the data has a changing TBP—a fact we have already noted in discussing Eq. (2.5).

Airplane vehicle. Suppose now that the vehicle is an airplane flying at constant velocity in a straight line at an altitude h. Then the plane of scatterers is now parallel to the axis of the cones (i.e., vehicle velocity vector), but offset: the intersections are now hyperbolas. The situation is shown in Fig. 2.6.

We see that the zero Doppler line is the only straight line: it is still normal to the vehicle velocity vector, and it passes through the vertical projection point of the vehicle. Hence, it is still possible to align the azimuth boresight with

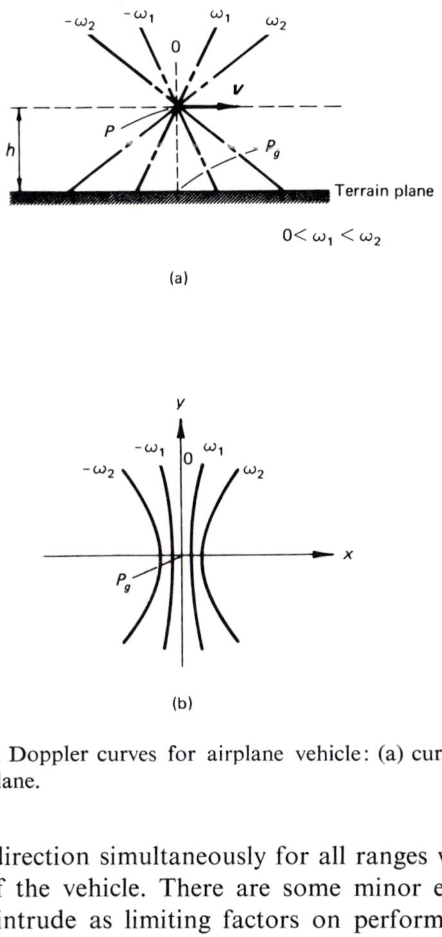

FIG. 2.6 Constant Doppler curves for airplane vehicle: (a) curves in vertical plane; (b) curves in terrain plane.

the zero Doppler direction simultaneously for all ranges with, e.g., a simple yaw positioning of the vehicle. There are some minor effects to be noted because they can intrude as limiting factors on performance. First, bandwidth is a function of range, therefore also azimuth resolution and required sampling frequency. Second, the bandwidths of selected signals depend on the boresight direction because, unlike the very low altitude case, the curves of constant Doppler are not equally separated. Third, the center, or carrier, frequency of the signals depends on range unless the antenna boresight coincides with the zero Doppler direction. This last observation will be seen to be important when we consider means of artificially achieving antenna boresighting—or more precisely, achieve correct offset, or carrier frequencies for all signals in the subsequent processing operations (Sec. 10.4).

Rotating earth. We must next take into account the motion of the plane of scatterers. For example, we assume that the area of the earth illuminated at any one instant is approximable by a plane; this is sufficiently accurate for most purposes. Then the rotation of the earth imparts a uniform relative velocity to all scatterers. This velocity we can in turn attribute to the vehicle itself. All this means is that the resultant velocity vector, $\mathbf{v}_r$—which determines the axis of the constant Doppler cones—is now at an angle to the vehicle velocity vector $\mathbf{v}$. Since we have been defining the x-coordinate of the plane of scatterers to be parallel to the satellite velocity vector (so that we can relate the fixed satellite attitude coordinates and the plane of scatterers, of course) this means simply that the family of curves of constant Doppler is rotated in the (x, R)-plane, as shown in Fig. 2.7.

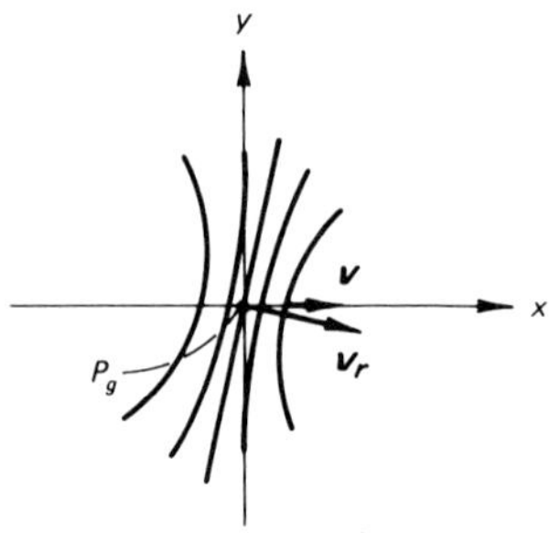

FIG. 2.7 Curves of constant Doppler in the terrain plane: vehicle above terrain plane and the latter in uniform motion.

All our observations of the "airplane case" remain true, if we just take into account the new yaw angle of the zero Doppler curve—which is still a straight line. In fact, this description suffices to describe satellite systems with near-circular orbits and not too extreme a range coverage. Of course, the relative velocity vector of the plane of scatterers is a function of latitude, among other things, so the rotation of the zero Doppler line is a function of latitude: this is an effect that must be corrected for (Chapter X).

Vehicle with eccentric orbit. To complete the description we consider eccentric orbits—that is, we assume that there is a nonzero component of vehicle velocity along the local vertical. This means that the axis of the cones of constant Doppler is no longer parallel to the plane of scatterers. The intersections of the plane of scatterers with the cones of interest remain hyperbolas. The situation is shown in Fig. 2.8.

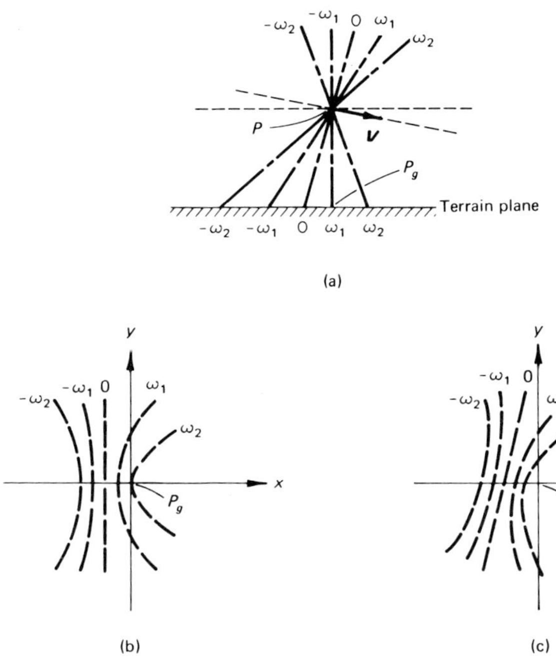

FIG. 2.8 Curves of constant Doppler for (e.g.) an orbital vehicle in an eccentric orbit in: (a) vertical plane—vehicle velocity vector **v** not normal to local vertical; (b) terrain plane for a stationary earth; (c) terrain plane for a rotating earth.

We note in particular that it will take a combination of pitch and yaw adjustments to realign the beam in the zero Doppler direction. *The zero Doppler line is a straight line, and it is clear by now that it always will be, as long as the scatterers lie in a plane with equal velocities*: the intersection of this plane with the zero Doppler plane must always be a straight line.

Scatterer motion. The movement of scatterers causes important effects. Not only may motion occur in the usual way—an object moves over the surface being mapped—but also, e.g., the curvature of the surface of a rotating planet causes each point on the surface to have a different relative velocity vector. The effects offer opportunities both for detecting and estimating the velocities of moving targets and limitations on the extent of the surface that can be satisfactorily imaged. We are here concerned with the motion effects on azimuth behavior.

Having already seen qualitatively what the Doppler space looks like for various relative velocities, we consider for concreteness a scatterer moving over a surface that is being mapped: we therefore want to compare the two appropriate Doppler spaces to see the effects. We assume, as is generally true, that the velocity vectors are constant during the time required to receive the azimuth history.

The significant components of the scatterer's velocity vector are the projection along the direction of the vehicle's velocity vector and the projection along the direction to the vehicle from the scatterer (the radial component), denoting their magnitudes v_P and v_R, respectively.

The effect of v_P is easily seen: it either makes the lines of constant Doppler closer (*case a*: if $\mathbf{v}_p$ is directed opposite to $\mathbf{v}$) or more separated (*case b*: if $\mathbf{v}_p$ is directed along $\mathbf{v}$). There are therefore two major effects:

1. The bandwidth of the azimuth history is greater (*case a*) or smaller (*case b*) than that of the stationary scatterers.
2. The duration of the azimuth history is smaller (*case a*) or greater (*case b*) than that of the stationary scatterers.

Thus the azimuth linear FM rate $k \equiv \Omega/T$ is greater (*case a*) or smaller (*case b*) than that of the stationary scatterers. In either case, the moving scatterer's azimuth history is mismatched to the processor set up for the stationary scatterers, and, thus, it will not compress as well. In order that it compress satisfactorily we require (see Sec. 6.2.1) that the quadratic phase error be less than $\pi/2$ rad: this leads to the requirement

$$\frac{v_P}{v} < \frac{2\pi}{\Omega T}.$$

For example, in a satellite situation $2\pi/\Omega T \approx 10^{-3}$, $v \approx 18{,}000$ mph, and so the condition is $v_P < 18$ mph.

The effect of v_R is easily seen: the lines of constant Doppler are rotated, the direction depending on the vehicle velocity vector and whether the scatterer's radial motion is opening or closing. There are two major effects:

3. The azimuth history passes through zero Doppler frequency either earlier or later than that of a stationary scatterer.
4. The azimuth history is not symmetrical about zero Doppler.

The first effect means that the azimuth position of the compressed response will be displaced from its true azimuth position. We can see this from the discussion in Sec. 2.6 of properties of linear FM signals. More fundamentally, we note that the synthetic aperture radar finds azimuth location via the Doppler coordinate—and the scatterer's motion belies this relation. It is easy to see from the discussion of Sec. 2.6 that the azimuth position error caused by the radial velocity v_R is

$$x_\Delta = \left(\frac{v_R}{v}\right) R_0 .$$

This is a perhaps surprisingly large error. In azimuth resolution units it is

$$\frac{x_\Delta}{\rho_a} = \left(\frac{v_R}{v}\right)\frac{2R_o}{D_h}.$$

R_o/D_h is generally a very large number: therefore, unless v_R/v is extremely small, the error will be many resolution units.

Example. Suppose $v = 25{,}000$ ft/sec and $R_o = 150$ naut mi, a possible pair of orbital parameters (approximately). Then, $x_\Delta = 36\ v_R$ ft, if v_R has units ft/sec!

The second effect means that the processor will probably discard some of the energy of the response. By the same token, a specially designed processor might be able to detect the presence of moving targets. This is a subject of prolonged investigation and generally will not be discussed here. We note in passing that we must have the fourth effect dominate the first effect.

2.5. THE AZIMUTH MODULATION

We now examine in more detail the Doppler modulation generated in the SAR configuration. This will provide a more general setting for the special case of quadratic phase modulation and, therefore, lead us to specifying more precisely the range of usefulness of the SAR system.

By ray optics, the phase of the received signal scattered by a point reflector has a phase modulation

$$\varphi(t) = 2k(R_o^2 + v^2t^2)^{1/2}, \qquad -\infty < t < \infty, \tag{2.14}$$

where the "wavenumber" $k \equiv 2\pi/\lambda$, λ is the wavelength of the radiated wave, and R_o is the range to the scatterer at $t = 0$: this phase modulation is due to the Doppler resulting from vehicle motion. φ' is shown in Fig. 2.9. The received signal has an amplitude modulation that has one factor equal to R_t^{-2},

$$R_t^2 = R_o^2 + (vt)^2, \qquad -\infty < t < \infty,$$

due, of course, to the changing range, and another factor—assumed real—equal to the two-way antenna pattern in the azimuth direction. Thus, for present purposes, we may take the azimuth modulation to be, in complex representation,†

$$A(t) = \frac{K}{R_o^2 + (vt)^2}\,|A(t)|\,\exp[i2k(R_o^2 + v^2t^2)^{1/2}], \qquad -\infty < t < \infty. \tag{2.15}$$

† The real signal would be $\mathrm{Re}\{A(t)\exp i\omega_0 t\}$ where ω_0 is the carrier frequency—assumed greater than half the bandwidth of A.

We shall assume that the signal is received in additive, white normal noise, and that we use its matched filter as the processor. We wish to investigate the intrinsic resolution of this signal (at the output of the matched filter) as a function of the duration T of the actually received data $A(t)$, $\frac{1}{2}T \leq t \leq \frac{1}{2}T$.

We have already observed that in the standard SAR configuration the signal duration T is determined by a fixed antenna pointed in azimuth

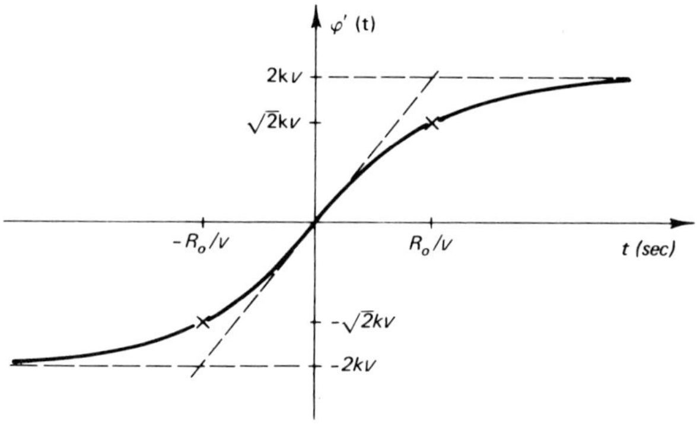

FIG. 2.9 The instantaneous frequency of the azimuth history for a synthetic aperture radar.

orthogonal (or nearly so) to the vehicle velocity vector. This is not the only possibility. If one wished to image a suitably confined region of terrain, one could "track" this region by steering, in azimuth, the antenna, or scanning, beam: one would intuitively feel—and correctly so, as we shall immediately see—that the resultant increase in data should produce a better image than a system using a fixed beam. In fact, such a scheme has been proposed as part of a planetary mapping system.

Modulation bandwidth. It is apparently not possible to find the spectrum $\tilde{A}$ of A—that is, the Fourier transform of Eq. (2.15)—exactly. (Even if we could, we would probably have to approximate the result to extract the gross features of interest.) But we can perform the calculation approximately sufficiently well for our purposes.

It is natural to consider first the situation $T \ll R_o/v$ where φ can be approximated by retaining only up to the quadratic terms of its Taylor series expansion; for if we also approximate the far-field antenna pattern by a Gaussian shaped function [and take the amplitude factor $(R_o + vt)^{-2} \approx R_o^{-2}$] we have a well-known integral. The calculation is given in App. 2A; the result is

$$\tilde{A}(\omega) \approx \mathcal{C} \exp\left[-\left(\frac{1}{T^2} - i\,\frac{2kv^2}{R_o}\right)^{-1} \cdot \frac{1}{2}\,\omega^2\right].$$

After passage through the matched filter, the signal has spectrum

$$|\tilde{A}(\omega)|^2 \approx \mathbb{C} \exp\left[-\frac{2T^2}{1+\left(\dfrac{2kv^2T^2}{R_o}\right)^2} \cdot \frac{1}{2}\omega^2 \right].$$

In fact, this form also is easily (inverse) Fourier transformed, so we have as the output of the matched filter

$$A_o(t) = \mathbb{C} \exp\left\{ -\left[\frac{2T^2}{1+\left(\dfrac{2kv^2T^2}{R_o}\right)^2} \right]^{-1} \cdot \frac{1}{2} t^2 \right\}, \qquad T \ll \frac{R_o}{v}.$$

It turns out that it is also natural to consider the situation

$$T \gg (R_o/v)/(kR_o)^{1/2},$$

then we can use the method of stationary phase (see App. 2A) to calculate approximately the spectrum of the modulation. Since always in SAR sytems $(kR_o)^{1/2}$ is an enormously large number, we are in the fortunate circumstance of having our two special cases overlap in their T-range of applicability. The explanation of the method of stationary phase and its application here is relegated to App. 2A; the result we principally want here is that the modulation bandwidth is given approximately by

$$\Omega \equiv \frac{4kv}{\left[1+\left(\dfrac{R_o}{vT}\right)^2\right]^{1/2}}, \qquad kv^2T^2 \gg 1. \tag{2.16}$$

Resolution. In the case $T \ll R_o/v$ we have calculated the approximate output to be a Gaussian-shaped pulse of standard deviation—which we take as a natural measure of resolution—

$$\rho_a \approx \frac{\sqrt{2}\,T}{\left[1+\left(\dfrac{2kv^2T^2}{R_o}\right)^2\right]^{1/2}}, \qquad T \ll \frac{R_o}{v}. \tag{2.17a}$$

In the case $kv^2T^2/2R_o \gg 1$, we have found the approximate bandwidth Ω; defining resolution to be $2\pi/\Omega$ we have

$$\rho_a \approx \frac{2\pi}{4kv}\left[1+\left(\frac{R_o}{vT}\right)^2\right]^{1/2}, \qquad \frac{kv^2T^2}{2R_o} \gg 1. \tag{2.17b}$$

The resolution, as given by these two equations, is sketched in Fig. 2.10. We note the T-region where T-regions T_1 and T_2 (defined in Fig. 2.10) intersect: this is the usual operating T-region for SAR systems. By definition there, both the large time bandwidth product and quadratic phase approximation conditions are satisfied.

In the T-region near zero, the resolution goes to zero: this is the "brute force" region where resolution is directly determined by physical antenna beamwidth.

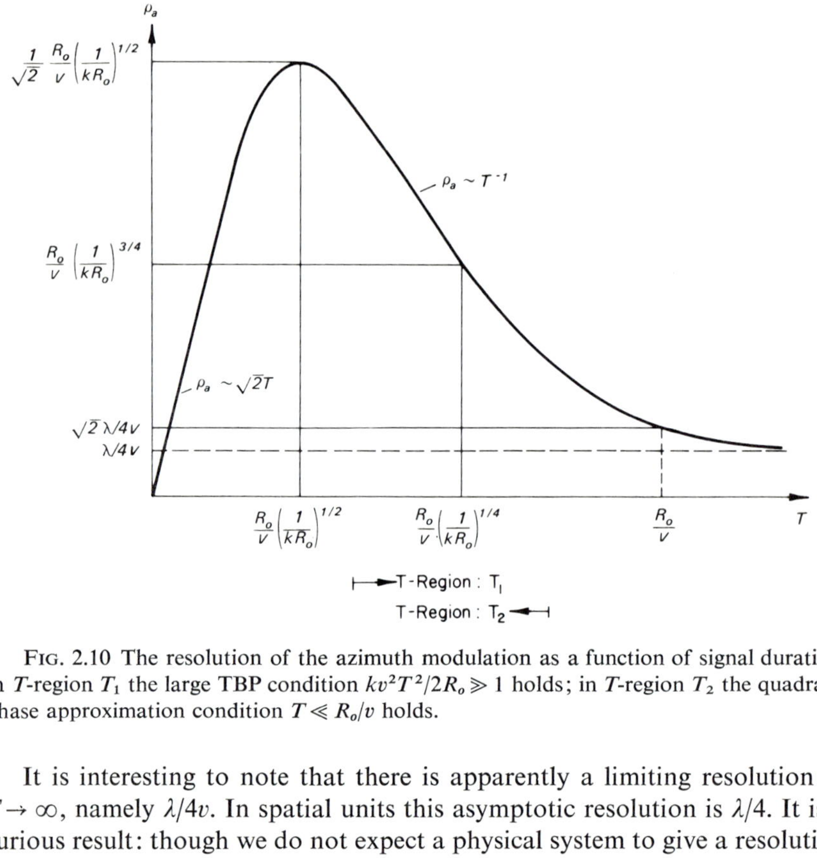

FIG. 2.10 The resolution of the azimuth modulation as a function of signal duration. In T-region T_1 the large TBP condition $kv^2T^2/2R_o \gg 1$ holds; in T-region T_2 the quadratic phase approximation condition $T \ll R_o/v$ holds.

It is interesting to note that there is apparently a limiting resolution as $T \to \infty$, namely $\lambda/4v$. In spatial units this asymptotic resolution is $\lambda/4$. It is a curious result: though we do not expect a physical system to give a resolution better than the order of the sensing wavelength, the analytic apparatus that we used was simply physical optics! This limiting resolution is shown graphically in Fig. 2.11.

Time-bandwidth product (*TBP*). An important parameter of a signal is its TBP, a measure of the complexity of the signal processing equipment that will be needed. E.g., suppose we wished to numerically compress a linear FM signal using a digital computer: we would have to sample the signal at least at intervals of the reciprocal of the bandwidth and for the entire time

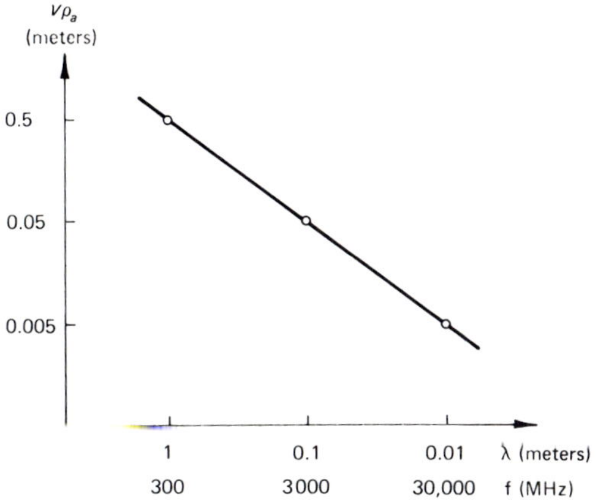

FIG. 2.11 Approximate ultimate resolution attainable from entire azimuth history.

extent of the signal. Similar remarks can be made for optical—or indeed any—signal processors.

We have

$$\mathrm{TBP} = \begin{cases} \dfrac{1}{2\pi}\left[1 + \left(\dfrac{2kv^2T^2}{R_o}\right)^2\right]^{1/2}, & T \ll (R_o/v), \\ \dfrac{1}{2}\dfrac{4kvT}{[1 + (R_o/vT)^2]^{1/2}}, & T \gg \dfrac{(R_o/v)}{(kR_o)^{1/2}}, \end{cases} \tag{2.18}$$

and this is plotted in Fig. 2.12 as a function of T. As we expect, for very

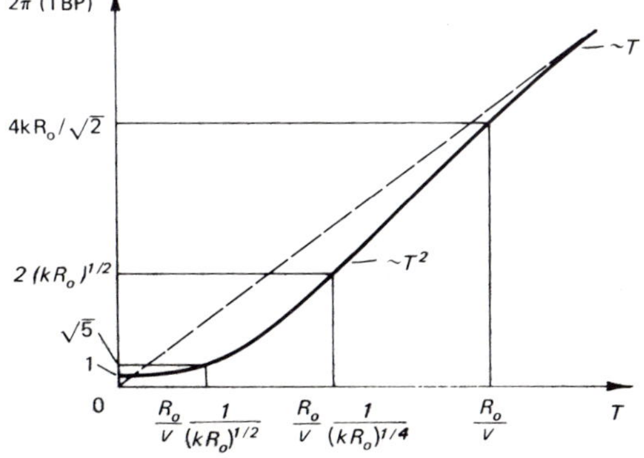

FIG. 2.12 The time-bandwidth product (TBP) of the azimuth modulation as a function of its length T.

small T the TBP is a constant of the order of unity—the exact constant depends on definitions. It then increases quadratically as T exceeds $(R_o/v)/(kR_o)^{1/2}$, a property of a linear FM signal. Ultimately as T exceeds R_o/v, the TBP must become a linear function of T: little further bandwidth can be gained.

Comparing Figs. 2.10 and 2.12 we see that when T is of the order of R_o/v and larger, ρ_a no longer decreases rapidly, but the TBP continues to increase. Therefore, unless other considerations dominate, on the basis of resolution and time-bandwidth product, *values of T of the order of R_o/v and greater are not desirable.*

2.5.1 Fixed Antenna

The above results will now be specialized to the case of a fixed antenna, of aperture D_h, pointed along the zero-Doppler direction. We shall use the approximation $T = R_o \lambda / v D_h$ to rewrite the above results; that is, we shall assume that the approximate beamwidth of the antenna is given by λ / D_h.

We observe that the condition $T \ll R_o/v$ now becomes $D_h \gg \lambda$. The condition $T \gg (R_o/v)/(kR_o)^{1/2}$ now becomes $D_h^2/\lambda R \ll \pi$: this is the well-known "near-field" condition for an aperture (see Sec. 5.1.1). The spatial resolution $v\rho_a$ and the time-bandwidth product, TBP, as functions of D_h, are plotted in Figs. 2.13 and 2.14, respectively.

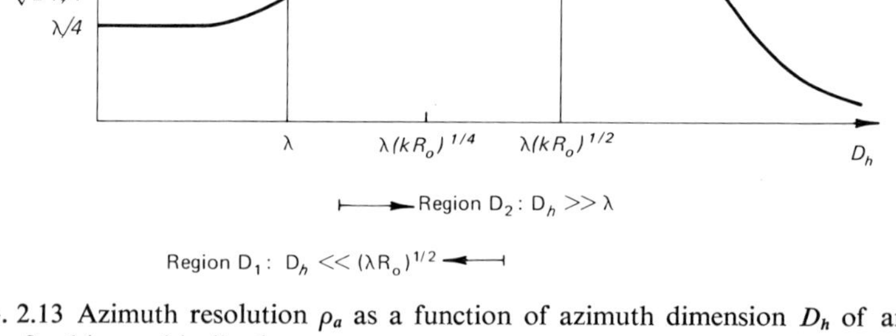

FIG. 2.13 Azimuth resolution ρ_a as a function of azimuth dimension D_h of antenna aperture, fixed beam, idealized pattern.

In the interesting range $D_h^2/\lambda R \ll \pi$, we have

$$v\rho_a \approx \frac{1}{4}\lambda\left[1 + \left(\frac{D_h}{\lambda}\right)^2\right]^{1/2}.$$

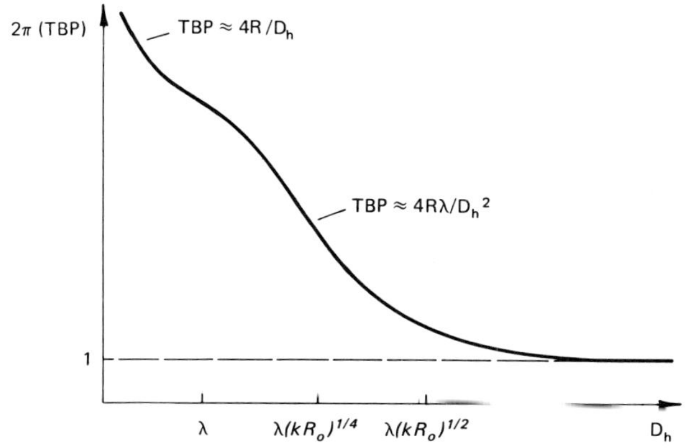

FIG. 2.14 TBP of azimuth signal as a function of azimuth aperture size (simplified far-field pattern).

The spatial resolution depends only on D_h and λ and is independent of v and R. If additionally $D_h \gg \lambda$, then we have

$$v\rho_a \approx \tfrac{1}{4}D_h.$$

The spatial resolution depends only on D_h and is independent of λ, v, and R.†

At the opposite extreme—the so-called "brute force" limit—we have, when $D_h^2/\lambda R_o \gg \pi$,

$$v\rho_a \approx \frac{R_o \lambda}{D_h}.$$

The spatial resolution depends, of course, on R_o, λ, and D_h.

2.6. THE LINEAR FREQUENCY MODULATION

A commonly used radar signal is a periodic sequence of pulses that are phase modulated with a quadratic function of time. The "instantaneous frequency" of the modulation, defined as the time derivative of the phase, is therefore a linear function of time. We shall refer to such a modulation as "linear FM"; it is also commonly termed a "chirp" pulse. Its use in radar arose because it provides a method of generating a pulse whose bandwidth is essentially independent of its envelope: thus high resolution, high energy signals can be designed under a peak power restraint.

† The reader may be bothered by the appearance of $\frac{1}{4}D_h$ instead of $\frac{1}{2}D_h$ [Eq. (2.71a)]: but in our approximations of this section we have not worried about factors of this size, being interested in the general ideas.

Such a signal can, of course, be of use in a synthetic aperture radar system. We have an even more important reason for studying it: the azimuth modulation, or "history," is, to a very good approximation, linear FM. Although such an approximation is not at all necessary to the synthetic aperture principle (Sec. 1.4), it provides an extremely convenient point of view in analyzing these systems.

A linear FM signal gives an example of a "large time-bandwidth product" signal. Such signals are often presently necessary in the practical implementation of these systems. Thus properties and insight relevant to such signals are obtained from the study of linear FM signals.

We begin with the now obvious definition.

Definition 2.1. A "linear FM signal" has a phase modulation that is a quadratic function of time and is a bandpass signal:

$$\begin{aligned} f(t) &\equiv F_m(t)\cos[\omega_o t + \tfrac{1}{2}kt^2] \\ &\equiv \mathrm{Re}\{F_m(t)\exp(i\omega_o t + \tfrac{1}{2}ikt^2)\}. \end{aligned}$$

F_m is the real envelope, ω_o the real carrier, or center, frequency, and k is the "linear FM rate." The "complex modulation" $F(t) \equiv F_m(t)\exp\tfrac{1}{2}ikt^2$.

The reason why we make the signal a bandpass signal will become clear when we consider the "compression," or matched filtering, of this type of signal below. Usually the envelope will be "pulse-like" centered on $t = 0$ and of some duration T, suitably defined.

Example 1. Let

$$F_m(t) = \mathrm{rect}\left(\frac{t}{T}\right) \equiv \begin{cases} 1, & |t| \le \tfrac{1}{2}T, \\ 0, & |t| > \tfrac{1}{2}T, \end{cases}$$

and

$$k \equiv \Delta/T.$$

Here, T is obviously the duration of F since f is identically zero when $|t| > \tfrac{1}{2}T$.

Example 2. Let

$$F_m(t) = \frac{1}{(2\pi T)^{1/2}}\exp\left(-\frac{1}{2}\frac{t^2}{T^2}\right).$$

Here, T is a measure of the duration defined as twice the radius of gyration of $|F|^2$.

A most important representation of a signal is by means of its Fourier transform, or spectrum. This, of course, is because of its use in characterizing its passage through linear devices; it is also useful in analyzing the effect of nonlinear operations on the signal. The spectrum we denote $\tilde{f}$:

$$\tilde{f}(\omega) = \int_{-\infty}^{\infty} e^{-i\omega t} f(t)\, dt.$$

Since the passage of bandpass signals through bandpass devices is best handled by use of the complex representation, we are also interested in the spectrum of the complex modulation

$$\tilde{F}(\omega) = \int_{-\infty}^{\infty} e^{-i\omega t} F(t)\, dt.$$

We approximate the spectrum of a linear FM signal by the method of stationary phase, discussed in App. 2A. The method is simpler for such a signal because the phase of the integrand is exactly a Taylor series up to quadratic terms: $I_2 = (-\infty, \infty)$. Thus:

Property 2.1. "If the magnitude F_m of a linear FM with FM rate k is approximately constant over intervals of length I_1 and $I_1 k^{1/2} \gg 1$, then

$$\tilde{F}(\omega) \approx \left(\frac{2\pi}{k}\right)^{1/2} F_m\left(\frac{\omega}{k}\right) \exp\left(-i\omega^2 2k + \tfrac{1}{4}i\pi\right)." \tag{2.19}$$

If we define $k \equiv \Delta/T$, where T is a measure of the duration of F_m, then as $I_1 \leq T$, we have

$$(T\Delta)^{1/2} \geq I_1 k^{1/2} \gg 1$$

for the necessary condition: $T\Delta$ is called the "time-bandwidth product," TBP, and it must be large in order to use the method of stationary phase. Further $F_m(\omega/k)$ then has approximate extent Δ, and so Δ becomes in this case a measure of the bandwidth of the linear FM.

The extreme usefulness of this theorem follows from the following observation. Suppose a signal is of very large time-bandwidth product (hereafter denoted TBP), and that it can be subdivided into pieces which are also of large TBP. Then each such piece contributes a certain part of the overall spectrum: we can relate uniquely each such piece of the signal with a piece of the spectrum. This is the basis for the intuitive notion that the instantaneous frequency can sometimes be used to describe the spectrum.

We now apply this theorem to an example and compare the approximation with the exact answer.

Example 3. Let

$$F(t) = \operatorname{rect}\left(\frac{t}{T}\right) \exp\left(\tfrac{1}{2}ikt^2\right), \qquad k \equiv \Delta/T.$$

We note that the instantaneous frequency is kt and thus varies linearly from $-\frac{1}{2}\Delta$ to $+\frac{1}{2}\Delta$ during the time F is nonzero. Here F_m is strictly a constant over T but is violently nonconstant at $\pm\frac{1}{2}T$. The condition

$$T_i\left(\frac{k}{\pi}\right)^{1/2} = \left(\frac{\Delta T_i}{\pi} \cdot \frac{T_i}{T}\right)^{1/2} \gg 1$$

leads us to the requirement $(\Delta T)^{1/2} \geq (\Delta T_i)^{1/2} \gg 1$; even so, we cannot expect the approximation to be good for ω/k near $\pm\frac{1}{2}T$, that is ω near $\pm\frac{1}{2}\Delta$. By the theorem we then have

$$\tilde{F}(\omega) \approx \left(\frac{2\pi T}{\Delta}\right)^{1/2} \operatorname{rect}\left(\frac{\omega}{\Delta}\right) \exp\left(\frac{-i\omega^2}{2k} + \tfrac{1}{4}i\pi\right). \tag{2.20}$$

The exact transform is, of course, the Fresnel integral, a tabulated integral. The calculation reveals that the phase and the amplitude factor $(2\pi T/\Delta)^{1/2}$ agree exactly. As we expect, the envelope differs, especially at $\omega = \pm\frac{1}{2}\Delta$. The exact magnitude is shown in Fig. 2.15.

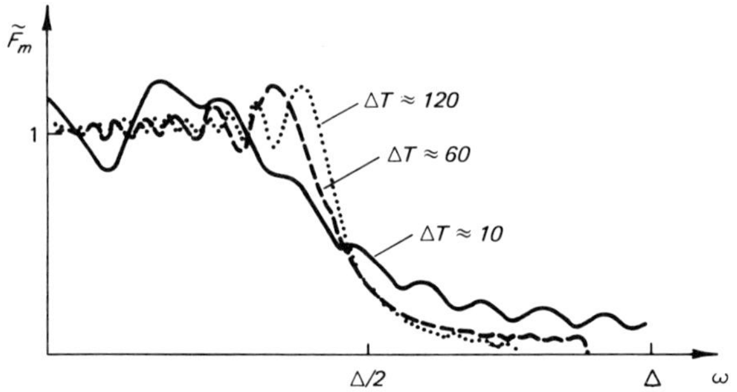

FIG. 2.15 The magnitude of the spectrum of a linear FM pulse of rectangular envelope of length T, and FM rate $k \equiv \Delta/T$; from Klauder *et al.* (1960).

From Fig. 2.15 we conclude that a time-bandwidth product ΔT of 100 or more is necessary for the approximation to be valid with this and similar discontinuous envelopes.

Matched filtering of linear FM signals An often used processing of signals involves matched filtering. For large TBP signals this operation on the signal is also often referred to as "compression." For many such signals, including the linear FM signal, we shall see there is a simple, approximate way to describe this operation.

If the complex modulation is again denoted

$$F(t) = F_m(t)\exp(\tfrac{1}{2}ikt^2)$$

then the complex representation of the matched filter is

$$\bar{F}(-t) = F_m(-t)\exp(-\tfrac{1}{2}ikt^2)$$

With the large TBP assumption, the spectrum of the complex modulation is

$$\tilde{F}(\omega) \approx \left(\frac{2\pi}{k}\right)^{1/2} F_m\left(\frac{\omega}{k}\right)\exp\left(\frac{-i\omega^2}{2k} + \frac{1}{4}i\pi\right).$$

The spectrum of the complex representation of the signal after the matched filtering operation is thus

$$\tilde{G}(\omega) \equiv |\tilde{F}(\omega)|^2 = \frac{2\pi}{k} \cdot F_m{}^2\left(\frac{\omega}{k}\right). \tag{2.21}$$

Thus the output signal is

$$G(t) = \frac{1}{2\pi}\int e^{i\omega t}\frac{2\pi}{k}F_m{}^2\left(\frac{\omega}{k}\right)d\omega = \int e^{-i(kt)\omega}F_m{}^2(\omega)\,d\omega, \tag{2.22}$$

and so we have:

Property 2.2. "The output of the matched filter of large time-bandwidth product linear FM signals is the Fourier transform of the squared envelope of the complex modulation suitably scaled."

Example 1. Let $F_m(t) = \text{rect}(t/T)$, so that $F_m{}^2(\omega/k) = \text{rect}(\omega/\Delta)$; then

$$G(t) = T\,\frac{\sin\frac{1}{2}\Delta t}{\frac{1}{2}\Delta t} \tag{2.23}$$

Normalizing G by the gain of the matched filter which is $(2\pi/k)^{1/2}$ over the signal band the ratio of the maximum of the output envelope squared to the maximum of the input envelope squared—which was unity—is

$$\frac{T^2}{2\pi/k} = \frac{\Delta T}{2\pi}.$$

The "compression" is the square root of the TBP, B in Hertz. Another measure of "compression" is the ratio of the extent of the signal before and after matched filtering. Before filtering it is reasonable to call the extent T; after filtering the signal is zero only outside an extent $2T$, but most of its energy is contained in an extent $2\pi/\Delta$—the width between nulls of the central lobe of G. Thus this ratio is $\Delta T/2\pi$ also.

There is a useful way to find gross properties of the output of the matched filter, especially when the input signal has been perturbed in various ways. Again the approximations make the method useful when the compression ratio, or TBP, is large. We observe that the phase of the spectrum of the matched filter has the derivative $-\omega/k$: this is the "delay" we can associate with a spectral region about ω. That is, we can subdivide the spectrum of the matched filter into segments centered on ω_n, $n = 0, 1, \cdots$. Each such segment has the associated delay $\omega_n/k = (\omega_n/\Delta)T$. Thus, using our knowledge obtained above about the relation between the envelope of signal and its spectrum, we see that the corresponding piece of the signal will be delayed $(\omega_n/\Delta)T$ sec by the action of the matched filter.

These observations lead to a simple graphical method of computing the gross features of the output. We first plot the relation between the instantaneous frequency and time (Fig. 2.16a). Then we plot the delay vs. frequency

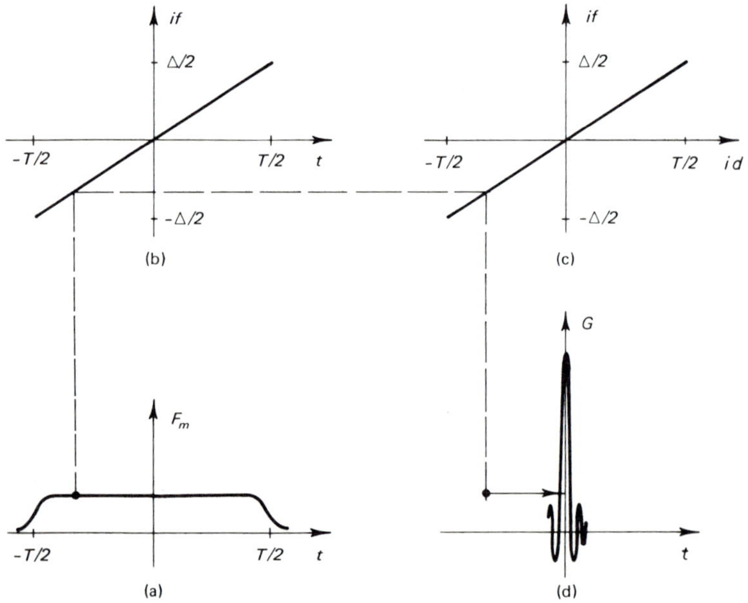

FIG. 2.16 Graphical method of computing some gross properties of the result of compression filtering a linear FM signal. (a) magnitude and (b) instantaneous frequency of linear FM signal; (c) the compression operation, and (d) the output signal.

of the matched filter (Fig. 2.16b). We can now approximately locate the output (Fig. 2.16c). The "fine structure" is given by Prop. 2.2.

From this graphical method it is easy to see that:

Property 2.3. "The approximate location in time of the output of the matched filter is the time of which the instantaneous frequency of the complex input signal is zero."

Paraphrased in terms of real signals, the theorem states that the approximate time of occurrence of the output signal is the time at which the instantaneous frequency of the input signal equals the center frequency of the matched filter. It is easy to see either analytically or from Fig. 2.16 that if ω_* is the frequency departure of signal carrier from the center frequency of the matched filter, then the time displacement is

$$T_d = T \cdot \frac{\omega_*}{\Delta}. \tag{2.24}$$

2.7. SAMPLING EFFECTS

The SAR system is normally operated with a periodic pulse modulation as we have described. The azimuth Doppler modulation is, therefore, sampled and must be recovered from these samples. As we shall see in detail in Chapter III, the sampled azimuth modulation received may be written

$$B_{hs}(x) \equiv B_h(x) \sum h_r(x - nX_s). \tag{2.25}$$

In this relation $B_h(x)$ is the azimuth complex modulation: its amplitude is determined by the geometry and far-field physical antenna pattern in azimuth, and its phase, $4\pi x^2/2\lambda R$, by the Doppler; also, h_r represents the recorder impulse response in the azimuth direction, and $X_s \equiv vT_s$ where $f_s \equiv 1/T_s$ is the sampling frequency (PRF) of the SAR system.

Since $\sum h_r(x - nX_s)$ is of period X_s in x, it has a Fourier series representation:

$$\sum h(x - nX_s) = \frac{1}{X_s} \sum \tilde{h}_r(nu_s) e^{inu_s x}, \qquad u_s \equiv \frac{2\pi}{X_s}.$$

The virtue of this representation is that it allows trivial calculation of the spectrum of the sampled signal:

$$\tilde{B}_{hs}(u) = \frac{1}{X_s} \sum \tilde{h}_r(nu_s) \tilde{B}_h(u - nu_s). \tag{2.26}$$

Thus, the spectrum of the sampled signal contains "aliases," $\{\tilde{B}_h(u - nu_s), n \neq 0\}$, of the spectrum $\tilde{B}_h$ of the original signal.

We have immediately a version of the famous Nyquist theorem:

Property 2.4. "If a signal has a spectrum zero for $|u| > \Omega$ and is sampled at a rate $u_s \geq \Omega$, then it can be reconstructed from the sample sequence by a linear, invariant filter."

In fact, from Eq. (2.26), we see that the linear invariant filter

$$h_s(u) \equiv \begin{cases} \dfrac{X_s}{\tilde{h}(0)}, & |u| \leq \frac{1}{2}u_s, \\ 0, & |u| > \frac{1}{2}u_s, \end{cases}$$

will pass only the term $n = 0$ with output spectrum $\tilde{B}_h(u)$; i.e., output $B_h(x)$.

The problem in many SAR systems, in particular those designed to operate near the "ambiguous limit" defined by Eq. (1.1), is that it is impossible to confine the far-field pattern of a finite physical aperture to a finite interval. Thus, the azimuth modulation B_h will contain frequencies too large to be sampled in a practical system: this is the "azimuth ambiguity problem" of SAR systems, for we shall now see that these aliases can cause false images.

These considerations are illustrated in Fig. 2.17.

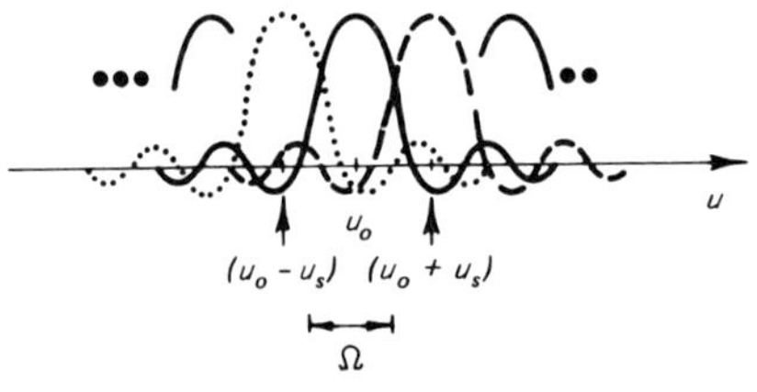

FIG. 2.17 The aliases due to sampling in a typical SAR system operating at the ambiguous limit.

Sampling of linear FM. We pursue further the sampling of the azimuth modulation by assuming that it is linear FM and further consider the effects on the output of the matched filter. There is, of course, nothing new to be said unless we also assume that the azimuth modulation cannot be sampled at a frequency sufficiently high to prevent overlapping of the aliases.

The primary effect, again, is that some aliases overlap into the passband of the matched filter. We use Prop. 2.3 to obtain their location after processing. These are false responses and are called "azimuth ambiguities" in synthetic

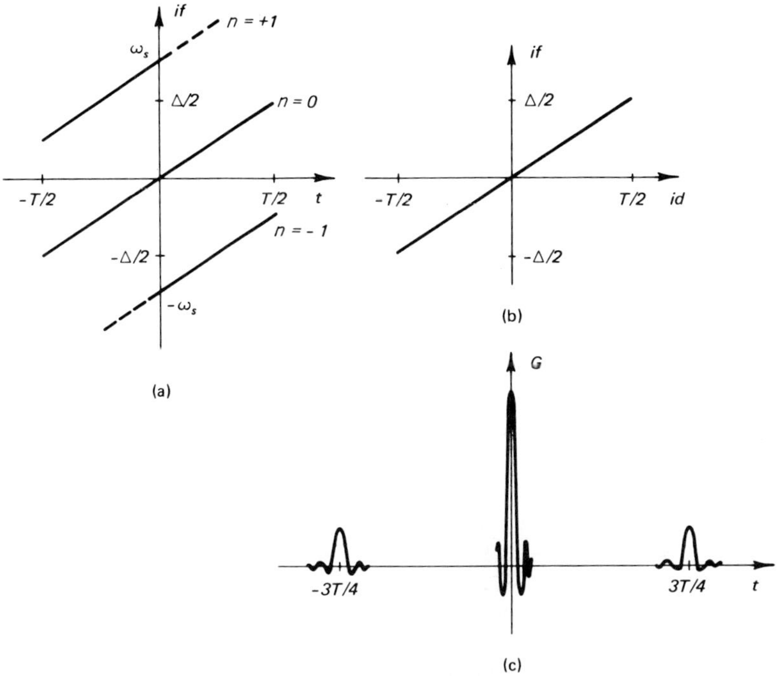

FIG. 2.18 The effect of undersampling the linear FM waveform of Fig. 2.17: (a) the instantaneous frequency vs. time curves for the aliases $n = 0, \pm 1$; (b) the compression filter operation; (c) the compressed output containing two false images.

aperture radar. The gross nature of the compressed output is calculated as described in Sec. 2.6 and is shown in Fig. 2.18.

There is another interesting case when the original signal is approximately bandpass, but the sampled signal is not: we shall encounter this case later. One must return to the real signal for the analysis, as the "negative spectra" must be handled. This situation, as shown in Fig. 2.19, is encountered in the "azimuth offset mode" of processing synthetic aperture radar data. The noteworthy additional effect is the presence of a "clutter."

Incorrect center frequency. If the sampled signal is on a center frequency somewhat different than the center frequency of the matched filter, then we know by Prop. 2.3 that the location of the compressed output will be displaced. This is the well-known effect of a Doppler velocity on a "chirp" range-finding radar. Generally, there will be other aliases now appearing in the passband of the processor, so false targets and clutter can occur. A typical situation is shown in Fig. 2.20.

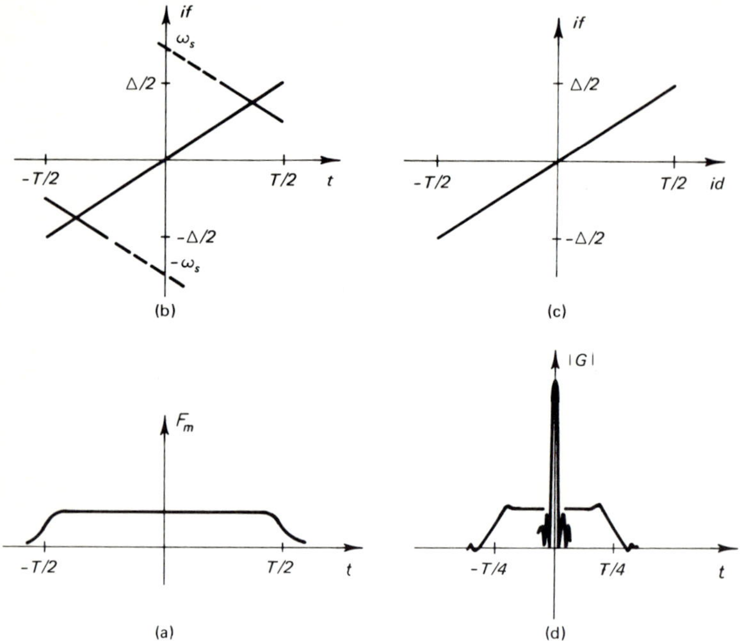

FIG. 2.19 The effect of undersampling a signal that is not narrow-band: (a) the envelope of linear FM signal; (b) the instantaneous frequency vs. time curves of the aliases; (c) the compression filter operation; (d) the partially compressed output containing clutter.

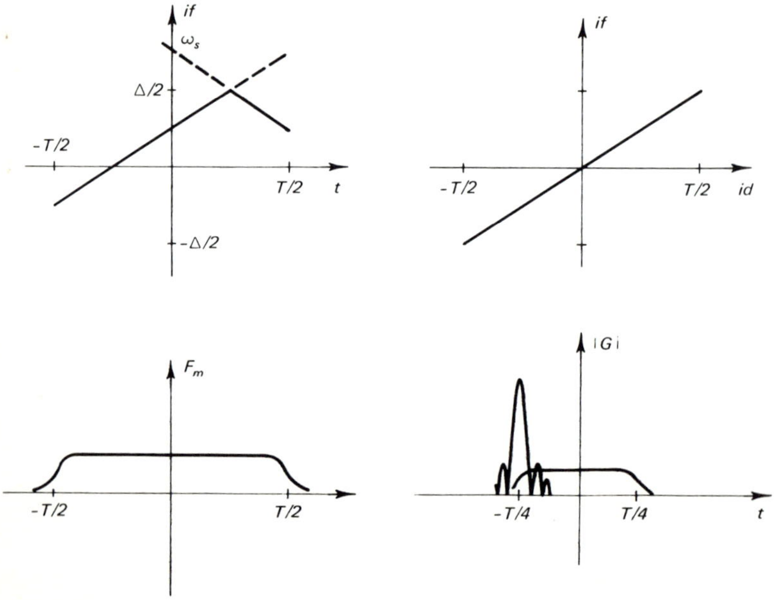

FIG. 2.20 The effect of sampling a linear FM signal on an incorrect center frequency.

2.8. DIGITAL PROCESSING

In the development thus far, we have derived the general requirements for a processor of the SAR data. Of particular importance was the bandwidth and time-bandwidth product. We now wish to consider the specific problem of "real-time" data processing—that is, processors which can produce an image of an object a very short time after its azimuth history has been gathered: this can be important in some possible applications of the SAR system. The coherent optical processors that we will discuss at length in Chapter V require exposure, development, and physical movement of photographic film or some similar operations: the time involved can be reduced to the order of a minute or so.

If we wish to do better than this, we might consider digital processing. Of course, we can be led to digital processing for other reasons: e.g., the wish to avoid coherent optical processing because of its special requirements of trained personnel, weight, volume, etc.

It will be helpful in the following to keep in mind that $D_h^2/\lambda R_m \ll 1$ in the SAR system.

Storage requirements. In order to compute the SAR image at a given range and azimuth, we must store, for each point sampled in range, the N samples required to adequately sample the azimuth history. The approximate time-bandwidth product is

$$\frac{\Omega T}{2\pi} = \frac{\lambda R_m}{\pi D_h^2} = N.$$

The number of range samples required is $M = \Delta R/\rho_a$ assuming the system has equal range and azimuth resolution. Thus, the number of samples that need to be stored simultaneously is

$$NM = \frac{1}{\pi}\frac{\lambda R_m}{D_h^2} M.$$

Recalling Eq. (1.1), for an SAR system mapping the entire unambiguous range interval, we have $M = \Delta R/\rho_a = v/2v$, so that

$$NM = \frac{\lambda R_m v}{2\pi D_h^2 v}.$$

The number of bits devoted to each sample depends upon the dynamic range of the data desired in the image. From 7 to 10 bits would be the normal range, often limited by the recording operation.

The processing operation. As we shall see in detail in Chapter IV, the optimum processing chosen for the SAR system ordinarily forms

$$|\mathscr{Z} * H|^2, \tag{2.27a}$$

where $\mathscr{Z}$ and H are the complex representations of the data and linear processor, respectively. For example, if the real data is

$$\mathscr{Z}(t) = |\mathscr{Z}(t)| \cos[\omega_o t + \theta(t)],$$

then the associated complex representation is $\mathscr{Z}(t) \equiv |\mathscr{Z}(t)| \exp i\theta(t)$ and $z(t) = \mathrm{Re}\{\mathscr{Z}(t) \exp i\omega_o t\}$. To see what this form is in terms of real signals and filters, we write out $\mathscr{Z}$ and H in terms of their real (e.g., $\mathscr{Z}_R$) and imaginary parts (e.g., $\mathscr{Z}_I$) and find the form

$$(\mathscr{Z}_R * H_R + \mathscr{Z}_I * H_I)^2 + (\mathscr{Z}_I * H_R - \mathscr{Z}_R * H_I)^2. \tag{2.27b}$$

The processing characterized by form (2.27) is shown in Fig. 2.21.

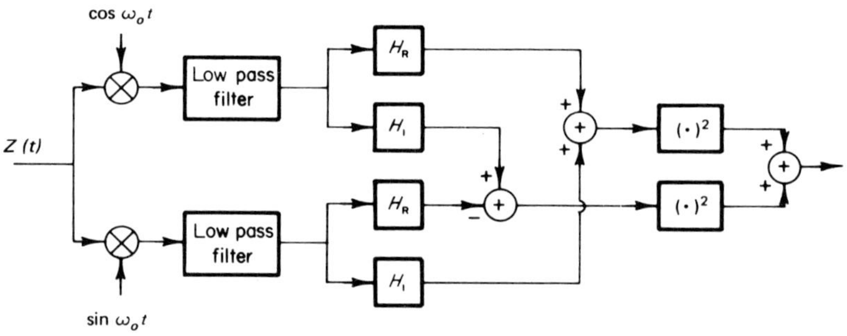

FIG. 2.21 The realization by a low pass system of the SAR processor.

We see that in order to get one azimuth output calculation, there must be $4N + 2$ multiplications and 3 additions, if N is the number of samples in the azimuth history. Since N is always large here, the processing time will be approximately $4NT_c$, where T_c is the time required for a multiplication.

Real-time processing. In order that unprocessed data not accumulate, the processor must perform the azimuth compression for all sampled ranges during the interpulse period T_s: this is what we shall mean by "real-time processing." Since the time required for this calculation is $4NMT_c$, we must have $4NMT_c \le T_s$, or,

$$T_c \le \left(\frac{D_h^2}{\lambda R_m}\right) \frac{D_h}{16v} \cdot \frac{1}{M}. \tag{2.28}$$

For an SAR system mapping the entire unambiguous range interval, condition (2.28) becomes

$$T_c \leq \left(\frac{D_h{}^2}{\lambda R_m}\right)\frac{D_h}{8v} \ll \frac{D_h}{8v}. \tag{2.28a}$$

Equation (2.28a) puts a formidable requirement on a processor: *the time allowable for a multiplication must be very small with respect to* 4×10^{-10} D_h *sec* (*units of* D_h *are meters*)!

We have assumed a strictly serial machine. Equation (2.28a) can be written in an amusing form. If we define $l_c \equiv vT_c$, we have necessarily

$$l_c \ll \tfrac{1}{8}D_h. \tag{2.28b}$$

During the time allowable for a multiplication, a light wave (*in free space*) *travels a distance very short with respect to the azimuth resolution!*

APPENDIX 2A: AZIMUTH MODULATION SPECTRUM

We wish to calculate the typical spectrum of the azimuth history in the general case. From ray optics we get the phase, and from simple free space propagation loss and an assumption of Gaussian shape for the two-way voltage pattern, we have the azimuth modulation due to a point scatterer to be

$$A(t) = \mathcal{C}\,\frac{\exp\left(-\dfrac{1}{2}\dfrac{t^2}{T^2}\right)}{R_o{}^2 + v^2t^2}\,e^{\varphi(t)} \tag{2A.1}$$

where

$$\varphi(t) \equiv 2k(R_o{}^2 + v^2t^2)^{1/2}.$$

The parameter T measures the extent (in temporal units) of the azimuth history as discussed in Sec. 2.4.

Limiting case $T \ll R_o/v$. If the spatial extent vT (in azimuth) of the terrain illuminated by the physical beam is small relative to the mean (say) range R_o to the terrain—i.e., if $T \ll R_o/v$—then the phase φ is well approximated by a quadratic phase. For, by the binomial expansion,

$$\begin{aligned}\varphi(t) &= 2kR_o\left[1 + \frac{1}{2}\left(\frac{vt}{R_o}\right)^2 + \cdots\right]\\ &\approx 2kR_o + \frac{k}{R_o}(vt)^2, \qquad |t| \ll \frac{R_o}{v}.\end{aligned}$$

The amplitude factor $(R_o + vt)^2 \approx R_o{}^2$ in this case.

Now the spectrum (Fourier transform) of the azimuth modulation is

$$\tilde{A}(\omega) = \mathcal{C} \int_{-\infty}^{\infty} e^{-i\omega t} \exp\left[-\frac{1}{2}\left(\frac{1}{T^2} - i\,\frac{2kv^2}{R_o} \right) \cdot t^2 \right] dt.$$

This is a well-known integral:

$$\tilde{A}(\omega) \approx \mathcal{C} \exp\left[-\frac{1}{2}\left(\frac{1}{T^2} - i\,\frac{2kv^2}{R_o} \right)^{-1} \cdot \omega^2 \right]. \qquad (2A.2)$$

The spectrum of the output of a filter of transfer function $\bar{\tilde{A}}(\omega)$—the matched filter—is therefore

$$|\tilde{A}(\omega)|^2 \approx \mathcal{C} \exp - \frac{1}{2}\left[\frac{2T^2}{1 + \left(\dfrac{2kv^2T^2}{R_o} \right)^2} \cdot \omega^2 \right]. \qquad (2A.3)$$

The output signal itself, $G(t)$, is then the (inverse) Fourier transform of Eq. (2A.3); this also is a well-known integral:

$$G(t) \approx \mathcal{C} \exp\left\{ -\frac{1}{2}\left[\frac{2T^2}{1 + \left(\dfrac{2kv^2T^2}{R_o} \right)^2} \right]^{-1} \cdot t^2 \right\}, \qquad T \ll \frac{R_o}{v}. \qquad (2A.4)$$

Large time-bandwidth product case, $kv^2T^2/R_o \gg 1$. When we do not have the quadratic approximation for φ we must try to find some other approximate method to get the results we want. It turns out that the method of stationary phase works quite well for the parameters of interest in SAR systems. We shall find that this method works well provided that

$$\frac{kv^2T^2}{2R_o} \gg 1. \qquad (2A.5)$$

We have already solved the situation when $vT/R_o \ll 1$; if we write condition (2A.5) in the form

$$kR_o\left(\frac{vT}{R_o} \right)^2 \gg 1,$$

we note that $kR_o = 2\pi R_o/\lambda$ is normally an extremely large number: therefore, our two approximate approaches will normally overlap in their range of application, and we will have the entire problem approximately solved.

The method of stationary phase can be motivated as follows. Given an integral with a complex phase in the integrand, we expand its phase in a Taylor series about the point (t_*, say) where its first derivative is stationary.

There will be an interval length, I_1 say, where the series to second-order terms will be a good approximation to the entire Taylor series. There is also an interval length, I_2 say, over which the magnitude of the integrand is approximately constant. Then we can write the integral in the form

$$\mathcal{C} \int_I \exp(i\alpha t^2)\, dt + \text{remainder},$$

where α is the coefficient of $(t - t_*)^2$ of the Taylor series of φ and $I \equiv \min(I_1, I_2)$. Now, referring to any tabulation of Fresnel integrals (e.g., Jahnke and Emde, 1945, p. 37) we see that if $I(\alpha)^{1/2} \gg 1$ then the Fresnel integral has achieved approximately its asymptotic value for $I(\alpha)^{1/2} \to \infty$, a constant. In conclusion, if I_1 and I_2 are defined as above, and t_* is defined to be the stationary point $[t \ni \varphi'(t_*) = 0]$, and $\min(I_1, I_2) \cdot [\varphi''(t_*)] \gg 1$, then

$$\int f_m(t) e^{i\varphi(t)}\, dt \approx \mathcal{C} \frac{f_m(t_*)}{[\varphi''(t_*)]^{1/2}} e^{i\varphi(t_*)}.$$

Applying this approximate method to find the spectrum

$$\tilde{A}(\omega) = \mathcal{C} \int e^{-i\omega t} A(t)\, dt$$

of $A(t)$ as given by Eq. (2A.1), we first find the stationary point by setting $d[-\omega t + \varphi(t)]/dt = 0$: if $|\omega| < 2kv$,

$$t_* = \frac{R_o}{v} \frac{-\omega}{(4k^2v^2 - \omega^2)^{1/2}}.$$

If $|\omega| > 2kv$, there is no solution. Thus, within this approximate method, the spectrum is zero for $|\omega| > 2kv$. Now the phase is well approximated about any point t_* by a quadratic over an interval $I_1 \ll R_o/v$—as we have already observed. The amplitude of A is well approximated by a constant over intervals $I_2 \ll \min(T, R_o/v)$. The quadratic coefficient $\varphi''(t_*)$ is $(\sqrt{2}vk^2R_o)^{-1} \times (4k^2v^2 - \omega^2)^{3/2}$. Therefore, if

$$\min\left(T, \frac{R_o}{v}\right) \cdot \left(\frac{1}{\sqrt{2}\, vR_o k^2}\right)^{1/2} \left(4k^2v^2 - \omega^2\right)^{3/4} \gg 1, \qquad |\omega| < 2kv, \quad (2A.5a)$$

then

$$\tilde{A}(\omega) \approx \mathcal{C}(4k^2v^2 - \omega^2)^{1/4} \cdot \exp\left\{-\frac{R_o{}^2}{2v^2T^2} \cdot \frac{\omega^2}{4k^2v^2 - \omega^2}\right\}$$

$$\exp\left\{i\frac{R_o}{v} \cdot \frac{\omega^2 + 4k^2v^2}{(4k^2v^2 - \omega^2)^{1/2}} - \tfrac{1}{4}i\pi\right\}, \qquad |\omega| < 2kv, \quad (2A.6)$$

and zero otherwise.

Condition (2A.5a) may be simplified to the asserted condition (2A.5). For if $T < R_o/v$, it is easy algebra to obtain (2A.5); if $T > R_o/v$, then by simple algebra we are led to the condition $R_o/\lambda \gg 1$, which is always true.

Of course, condition (2A.5) cannot be satisfied for $|\omega|$ sufficiently close to $2kv$: there our approximate method fails. But it is intuitively clear that nothing surprising can happen there: the magnitude of the spectrum must fall to zero in the neighborhood of $\omega = 2kv$.

We note that the factor $(4k^2v^2 - \omega^2)^{1/4}$ is a fairly flat function relative to the exponential factor and may be neglected. To get a measure of bandwidth —and hence resolution—of the signal, let us arbitrarily, but reasonably, define the bandwidth to be $\Omega \equiv 2\omega_c$ where $|\tilde{A}(\omega_c)|^2 = |\tilde{A}(o)|^2/e$: then

$$\Omega_c \approx \frac{4kv}{\left[1 + \left(\dfrac{R_o}{vT}\right)^2\right]^{1/2}}, \qquad \frac{kv^2T^2}{2R_0} \gg 1. \tag{2A.7}$$

COMMENTS AND REFERENCES

System parameters. The relations among the principal parameters [e.g., Eqs. (2.4), (2.7), and (2.9–2.13)] have been known for many years. The most interesting relation, that $\rho_h = D_h/2$, was first reported in the literature by Cutrona *et al.* (1961).

Doppler induced geometry. The possible relation between Doppler frequency and spatial location has been exploited in other active sensing systems. In particular, the rotation of some planetary bodies has enabled imaging/mapping—i.e., radar astronomy; for a survey see Muehleman *et al.* (1965). The imaging of rotating objects with relation to the SAR technique has been discussed by Brown (1967).

Linear FM. About the first, and a good, discussion of the linear FM was given by Klauder *et al.* (1960). For a later and extensive discussion see Cook in Berkowitz (1965) and Cook and Bernfeld (1967).

Sampling. For a discussion of the sampling of signals see almost any book on systems or communication theory: e.g., Brown (1963b).

III

SAR Signal, Image, and SNR

3.1. INTRODUCTION

We have now described at some length the SAR principle, its typical configurations, the approximate relations among its principal parameters, and, in general, the geometry of the operation by which it forms an image of the terrain.

We as yet have not discussed "signal-to-noise" ratio, which is usually one of the primary concerns of radar analysis. By "noise" is customarily meant additive noise, often Gaussian and often "white," that is due to the unavoidable thermal motion of the material universe. But this type of noise is a problem, obviously, only if the signal is not sufficiently larger than the noise—or, in radar usually, only if the radiated power is not sufficiently large. In some applications of SAR systems there is no significant radiated power restraint: e.g., a relatively large aircraft vehicle, mapping an interval of terrain at relatively short range would usually allow the removal of any practical limitations on radiated power.

In fact, an SAR system with just such a vehicle and geometry is likely to be bothered most by *multiplicative* errors—phase errors—due to random buffeting by the often large turbulence of the lower atmosphere. Even if compensating corrections were sensed and introduced, we would expect there to be residual multiplicative errors that may well have more significance on system performance than the weak additive noise.

Nevertheless, we shall concern ourselves in this chapter with models that include only additive, white noise. This is for several reasons: (1) additive noise can indeed be the most significant cause of irreducible error in some SAR systems (e.g., those using an orbital vehicle); (2) it is the basis, implicitly or explicitly, of most of the analysis and design of these systems; (3) any of several criteria lead to similar processors which include the "matched filter" (Chapter IV) which performs the intuitively desirable azimuth compression operation; and (4) reasonable criteria based on models including multiplicative errors do not lead to similar processors, and some of them are very complicated (Chapter X). We shall, in fact, assume that the processor is a matched filter in the present chapter.

We begin with a careful derivation of the signal generated by the SAR system. We shall model the physical antenna system by an aperture with an illumination function, use the Kirchoff-Huygens' diffraction integral throughout, and model the scattering mechanism by a multiplicative "reflectivity density." It will be seen that the mapping from terrain (so modeled) to the two-dimensional recorded data is a linear, quasi-invariant operation. This allows qualified use of the response to a "point" scatterer to characterize the SAR system. The complicated details of the sampling and matched filtering operations are next discussed, leading to a natural, two-channel model of linear, invariant subsystems for periodic pulse modulations. Finally, the signal-to-noise ratio is calculated.

3.2. THE RECEIVED SIGNAL

We wish now to determine more carefully the analytic form of the signal that is generated by the SAR system. We do this in a more general way than is necessary for the purposes of the present chapter in order to have the result for later use.

The radiated far field. We begin by computing the scattered field for a radiated waveform with a sinusoidal time variation; later we sum the result over all the sinusoids—i.e., over the frequency spectrum—of the radiated

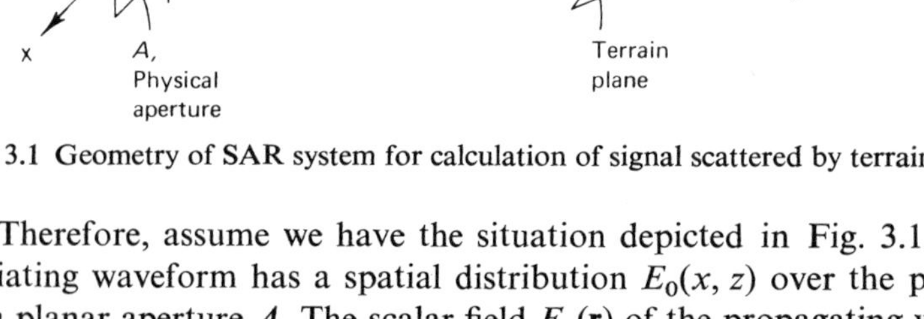

FIG. 3.1 Geometry of SAR system for calculation of signal scattered by terrain.

signal. Therefore, assume we have the situation depicted in Fig. 3.1 where the radiating waveform has a spatial distribution $E_0(x, z)$ over the physical antenna planar aperture A. The scalar field $E_f(\mathbf{r})$ of the propagating wave at any point **r** in space is given, sufficiently well for our purposes, by the Kirchoff-Huygens diffraction integral:

$$E_f(\mathbf{r}) = \frac{-ik}{4\pi} \iint_A \frac{e^{ik|\mathbf{r}-\mathbf{r}'|}}{|\mathbf{r}-\mathbf{r}'|} E_0(\mathbf{r}')\, dA'. \tag{3.1}$$

Here we have omitted writing the time dependence exp $i\omega t$, $k \equiv \omega/v$.

We have pointed out at some length above (especially Sec. 2.5) that normally small angles are of interest in the SAR system. We can then simplify Eq. (3.1) somewhat by using the approximation

$$|\mathbf{r}-\mathbf{r}'| \approx (y-y') + \frac{(x-x')^2+(z-z')^2}{2(y-y')}$$

in the phase and $|\mathbf{r}-\mathbf{r}'| \approx R_o$ in the amplitude of $(\exp ik|\mathbf{r}-\mathbf{r}'|)/|\mathbf{r}-\mathbf{r}'|$ of Eq. (3.1).

In particular, in the far field of the physical aperture, the phases $x'^2/2(y-y')$ and $z'^2/2(y-y')$ are negligible due to the very small relative size of the physical aperture; thus

$$E_f(\mathbf{r}) \approx \frac{-ik}{4\pi R_o} \exp\left[iky + \frac{ik}{2y}(x^2+z^2)\right]$$

$$\cdot \iint_A \exp\left[-i\frac{kx}{y}x' - i\frac{kz}{y}z'\right] E_0(x',z')\, dx'\, dz'.$$

We, of course, retain the quadratic terms in x^2 and z^2, representing our approximation to the spherical wavefront in the far field. We recognize the Fourier transform that appears in this relation, and so finally we may write

$$E_f(\mathbf{r}) \approx \frac{-ik}{4\pi R_o} \exp\left[iky + \frac{ik}{2y}(x^2+z^2)\right] \tilde{E}_0\left(\frac{k}{y}x, \frac{k}{y}z\right). \tag{3.2}$$

The scattered field. The field E_f illuminates the terrain and creates a field that propagates back to be intercepted by the physical aperture. On a surface S "near" the terrain we denote this field $E_{fs}(\mathbf{r})$ and we assume that

$$E_{fs} = L_\psi\{E_f\}, \qquad \mathbf{r} \in S, \tag{3.3}$$

where L_ψ is a linear operation. Generally L_ψ is very difficult to characterize. For present purposes we note that usually $E_{fs}(\mathbf{r}_1)$ depends upon $E_f(\mathbf{r}_2)$ only if $|\mathbf{r}_1 - \mathbf{r}_2|$ is of the order of a few wavelengths. If the SAR system resolution is much larger than the wavelength—as is usually true—then we make little error most of the time by assuming Eq. (3.3) has the special form

$$E_{fs}(\mathbf{r}) = \psi(\mathbf{r}; \ldots)E_f(\mathbf{r}), \qquad \mathbf{r} \in S. \tag{3.4}$$

Defining precisely the "field E_{fs} on the surface S" near the terrain seems necessarily to bring in the full complexity of the scattering problem which is, simply put, intractable. We shall, however, be able to say precisely what we mean by this for objects amenable to the method of physical optics (see App. 3A). But we know generally what we want S to be: it should follow the terrain geometry fluctuations of the order of the system resolution. Furthermore on S we ignore the radial component of the scattered field because it propagates negligibly to the far-field—the location of the physical receiving aperture. The definition of ψ, implicit in Eq. (3.4), can account to some extent for penetration, i.e., volume scattering.

Of course, effects such as multiple reflections over distances large relative to the resolution occasionally occur and are not modeled by Eq. (3.4). But, if Eq. (3.4) is not approximately true, then the usefulness of the SAR system as an imaging system would apparently be greatly reduced in almost any possible application.

The notation for the *reflectivity density*, $\psi(\mathbf{r}; \ldots)$, indicates that ψ depends on other factors, such as angles of incidence of the illumination and the wavelength. Indeed, the SAR system can be competitive in some applications only if ψ depends upon some parameter of interest in a peculiar way.† If ψ is defined relative to a fixed terrain coordinate system then $\psi(x - vt, y, z; \ldots)$ appears in Eq. (3.4) for the SAR system.

The scattered field at the aperture. The scattered field E_{fs} propagates back to the receiving aperture—where we denote it by E_s—in a manner that again can be calculated by the Kirchoff-Huygens diffraction integral. We can again use the quadratic approximation: for this case the quadratic phases $x^2/2y$ and $z^2/2y$ *outside* the integral can be neglected due to the small relative size of the physical aperture:

$$E_s(\bar{r}) \approx \left(\frac{-ik}{4\pi R_o}\right) \iint_S dS' \exp\left[iky' + \frac{ik(x'^2 + z'^2)}{2y'}\right]$$
$$\cdot E_{fs}(\bar{r}') \exp\left[-i\left(\frac{kx}{y'}\right)x' - i\left(\frac{kz}{y'}\right)z'\right], \tag{3.5}$$

where E_{fs} is given by Eqs. (3.2) and (3.4).

The received signal. The received wave passes through the aperture, receiving weighting $E_0(x, z)$, and is focused/integrated to give the received temporal signal $S_\omega(t)$. Because of the Fourier transform kernel exp $i(k/y')$

† As a curious but potentially important example, it has been observed that the image (in visible light) of some kinds of vegetation can reveal the presence of certain valuable minerals at depths up to tens of feet in the underlying soil.

$(-xx' - zz')$ appearing in the form of Eq. (3.5) for E_s, this operation introduces the far-field pattern $\tilde{E}_o$ again: we find

$$S_\omega(t) \approx \left(\frac{ik}{4\pi R_o}\right)^2 \iint_S ds' \exp\left[i2ky' + \frac{ikx'^2}{y'} + \frac{ikz'^2}{y'}\right]$$
$$\cdot \tilde{E}_0{}^2\left(\frac{k}{y'}x', \frac{k}{y'}z'\right)\psi(x' - vt, y', z'; \ldots). \tag{3.6}$$

The received signal is therefore

$$s_\omega(t) = \mathrm{Re}\{S_\omega(t)e^{i\omega t}\} \tag{3.6a}$$

when the radiated signal has temporal dependence exp $i\omega t$.

Modulated signal. The temporal variation of $S_\omega(t)$ is very slow relative to signal modulations of interest to us. If the signal actually radiated has temporal behavior $\mathrm{Re}\{F(t)\exp i\omega_c t\}$—we call F the complex signal modulation and ω_c the radiated carrier frequency—then the received signal is

$$s(t) = \mathrm{Re}\{S(t)e^{i\omega_c t}\} \tag{3.7a}$$

where

$$S(t) \equiv \frac{1}{2\pi}\int d\omega\, e^{i\omega t}\tilde{F}(\omega)S_{\omega+\omega_c}(t). \tag{3.7b}$$

We note that usually the only part of $S_{\omega+\omega c}$ that has significant k dependence is the phase $2ky'$ which accounts for the round trip delay time. In the other places where k appears, we usually can substitute $k_c \equiv \omega_c/v$. Note that in particular ψ depends on k, and the SAR system of course senses ψ as it appears to radiation of mean frequency ω_c. Thus to write it out,

$$S(t) \approx \left(\frac{ik}{4\pi R_o}\right)^2 \iint_S dS' F\left(t - \frac{2y'}{v}\right) \exp\left[\frac{ik}{y'}(x'^2 + z'^2)\right]$$
$$\cdot \tilde{E}_0{}^2\left(\frac{k}{y'}x', \frac{k}{y'}z'\right) \cdot \psi(x' - vt, y', z'; \lambda_c, \ldots). \tag{3.7c}$$

Terrain plane reference. The integral in Eq. (3.7c) is over the surface S whose reflectivity density we have denoted $\psi(\mathbf{r}; \ldots)$. It is convenient, and often meaningful, to assume S is a planar surface at or near the terrain—see Fig. 3.1. If the coordinates in this plane are (x, ζ) then we have $\lambda = R_m + \zeta \cos \gamma_o$ and $z = \zeta \sin \gamma_o$, where γ_o is the depression angle of the physical antenna's boresight. We denote ψ after these relations for y and z are substituted by $g(x - vt, \zeta; \ldots)$; further we assume that $\tilde{E}_0{}^2(x, z) = \tilde{E}_{0h}(x)\tilde{E}_{0v}(z)$† and define

$$B_h(x; \zeta) \equiv \exp\left[\frac{ikx^2}{R_m + \zeta \cos \gamma_0}\right] \tilde{E}_{0h}^2\left(\frac{k}{R_m} x\right) \tag{3.8a}$$

and

$$B_v(\zeta) \equiv \exp\left[\frac{ik\zeta^2 \sin^2 \gamma_0}{R_m + \zeta \cos \gamma_0}\right] \tilde{E}_{0v}^2\left(\frac{k \tan \gamma_0}{R_m} \zeta\right). \tag{3.8b}$$

Then we finally may write

$$S(t) = \left(\frac{ik}{4\pi R_m}\right)^2 \int dx \int d\zeta\, F\left(t - \tau_m - \frac{2\zeta \cos \gamma_o}{v}\right) \cdot B_v(\zeta) B_h(x - vt; \zeta) \cdot g(x, \zeta; \lambda_c, \ldots). \tag{3.9}$$

This is the temporal complex signal modulation received by the SAR system model we have developed here; the real signal is obtained via Eq. (3.7a).

Periodic modulation and two-dimensional recording. In SAR systems a periodic modulation of relatively short pulses is generally used. We therefore now set

$$F(t) = \sum F_0(t - nT_s),$$

where T_s is the period, the sampling frequency therefore $\omega_s \equiv 2\pi/T_s$, and $F_0(t)$ is centered on $t = 0$ and of duration very short relative to T_s. As we have discussed, each synchronized, T_s-length piece of data is recorded side by side to generate a two-dimensional format, after mixing the received signal down to an *offset frequency* ω_o. If we retain the same coordinates for simplicity we have the two-dimensional recorded signal

$$s(t, x) = \begin{cases} \mathrm{Re}\{S(t, x)e^{i\omega_o t}\}, & \text{range offset case,} \\ \mathrm{Re}\{S(t, x)e^{iu_o x}\}, & \text{azimuth offset case,} \end{cases} \tag{3.10a}$$

where

$$S(t, x) \equiv \left(\frac{ik}{4\pi R_o}\right)^2 \int dx' \int d\zeta F_0\left(t - \frac{2\zeta \cos \gamma_o}{v}\right) B_v(\zeta) \cdot \sum h_r(x - nX_s) B_h(x' - x; \zeta) \cdot g(x', \zeta; \ldots). \tag{3.10b}$$

Here $u_o \equiv \omega_o/v$ is called the *azimuth offset frequency*, ω_o the *range offset frequency*, $X_s \equiv vT_s$ is the spatial sampling distance, and h_r is the recorder impulse response in the azimuth direction. (The latter is carried along until it is

† This is true for all cases of interest to us here: see, e.g., Chapter VI. In particular, it is true whenever the aperture illumination $E_0(x, z)$ can be written in factored form.

convenient to discard it; its effect is easily accounted for—see Sec. 3.5.) The two offset frequency cases will be discussed below.†

If we neglect the presence of ambiguities by assuming that $B_h(x;\zeta)$ is adequately sampled then we can get a simpler form of Eq. (3.10) that is an extremely useful model for the SAR system. For then‡

$$\sum h_r(x - nx_s)B_h(x' - x;\zeta) \approx B_h(x' - x;\zeta)$$

and Eq. (3.10b) becomes

$$S(t,x) = \left(\frac{ik}{4\pi R_o}\right)^2 \int dx' \int d\zeta\, F_0\left(t - \frac{2\zeta\cos\gamma_o}{v}\right) \cdot B_v(\zeta)\cdot B_h(x' - x;\zeta)\cdot g(x',\zeta;\ldots). \tag{3.10c}$$

Equation (3.10c) characterizes the mapping from the terrain reflectivity g to the recorded two-dimensional data. It is a linear operator with kernel

$$K(t-\zeta, x-x';\zeta) \equiv \left(\frac{k}{4\pi R_o}\right)^2 F_0\left(t - \frac{2\zeta\cos\gamma_o}{v}\right) B_v(\zeta)B_h(x'-x;\zeta)$$

that is not solely a function of the differences $t - \zeta$ and $x - x'$, and therefore the operator is not (spatially) invariant. We know physically that generally it is not because, in simplest terms, the azimuth linear FM rate depends on range, or ζ.

However, it is true that normally the vertical far field pattern $|B_v(\zeta)|$ and the azimuth complex modulation vary very slowly with ζ compared to $F_0(t - 2\zeta\cos\gamma_0/v)$. Therefore, the linear operator defined by Eq. (3.10c) is "quasi-invariant" in the sense that, if the object is displaced by distances $\mathbf{r}_\Delta$ small with respect to the range interval and depth of focus of the SAR system —distances normally very large with respect to the system resolution—then the recorded data $s(t, x)$ remains approximately invariant except for being displaced by $\mathbf{r}_\Delta$.

Point scatterer. Suppose that $g(x,\zeta;\ldots) = g_o\delta(x - x_o, \zeta - \zeta_o)$, a "point" scattering object.§ Then, from Eq. (3.10b), we have

$$\begin{aligned} S(t,x) &= g_o\left(\frac{ik}{4\pi R_o}\right)^2 F_0\left(t - \frac{2\zeta_o\cos\gamma_o}{v}\right)\cdot B_v(\zeta_o)\cdot B_h(x - x_o;\zeta)\sum h_r(x - nX_s) \\ &\equiv K_o F_0(t - \tau_o)B_h(x - x_o;\tau_o)\sum h_r(x - nX_s), \end{aligned} \tag{3.11}$$

† We have derived Eq. (3.10) from Eq. (3.9) rather sketchily: the details are too arduous to set down here though they are interesting in some respects—perhaps only academically interesting, however.

‡ We assume that, as is done in practice, the recording operation h_r is designed so that the samples are smoothed as recorded; then $h_r(nu_s) \approx 0$, $n \neq 0$.

§ In Sec. 3.6 we discuss the relation between the scattering density, g—e.g., for a small object—and the conventionally defined far-field cross section.

where we have defined the delay

$$\tau_o \equiv \frac{2\zeta_o \cos\gamma_o}{v}, \qquad \text{and} \qquad K_o \equiv g_o\left[\frac{ik}{4\pi R_o}\right]^2 B_v(\zeta_o).$$

We note that the two-dimensional received signal scattered by a point object is a product of two functions, one a function of x and one a function of delay, except for the slow variation of B_h and B_v with τ.

3.3. SIGNAL SPECTRUM: POINT SCATTERER

The two dimensional signal of Eq. (3.11) is to be processed by the matched filter. As usual, we gain considerable insight if we consider this operation in the frequency, or transform, domain. For then the somewhat complicated azimuth modulation can be better understood, especially in the light of the discussions of Secs. 2.6 and 2.7.

Before we begin, we note that any practical processor can only handle a certain x-interval X of data. It is apparent how X is chosen on the basis of the above development: it will be the length of the azimuth history that is adequately sampled. We define the indicator of X,

$$I_X(x) \equiv \begin{cases} 1, & |x| < \frac{1}{2}X, \\ 0, & |x| > \frac{1}{2}X, \end{cases}$$

The two-dimensional transform of Eq. (3.11) reduces to two one-dimensional transforms due to the very slow τ_o-variation of $B_h(x - x_o;\, \tau_o)$ relative to that of $F(t - \tau_o)$. From Sec. 2.7 we have, if $\mathscr{F}$ denotes the Fourier transform,

$$\mathscr{F}\{B_h(x - x_o\,;\, \tau_o)\Sigma \tilde{h}_r(x - nX_s) \cdot I_X(x)\}$$
$$= \frac{1}{X_s}\Sigma \tilde{h}_r(nu_s)\mathscr{F}\{B_h(x - x_o\,;\, \tau_o)I_X(x)\}\Big|_{u-nus}$$

Further,

$$\mathscr{F}\{B_h(x - x_o\,;\, \tau_o)I_X(x)\} = e^{-ix_o u}\mathscr{F}\{B_h(x\,;\, \tau_o)I_X(x + x_o)\},$$

and we now use the large TBP property of B_h to write, by Prop. 2.1 of Sec. 2.6,

$$\mathscr{F}\{B_h(x\,;\, \tau_o)I_X(x + x_o)\} \approx \mathbb{C}\left|B_h\left(\frac{u}{k}\right)\right| I_X\left(\frac{u + x_o}{k}\right) \exp\left[\frac{-iu^2}{2k}\right].$$

Thus finally the spectrum of the sampled azimuth complex signal can be written down: it is

$$\Sigma B_n(u\,;\, x_o) \exp\left[\frac{i(u - nu_s)^2}{2k} + i(u - nu_s)x_o\right] \tag{3.12}$$

where

$$B_n(u;\, x_o) \equiv \left| B_h\!\left(\frac{u - nu_s}{k}\right)\right| I_X\!\left(\frac{u - nu_s + x_o}{k}\right)$$

and we have assumed that $h_r(nu_s)$ is approximately a constant for n of interest.

The two-dimensional spectrum of signal is, therefore,†

$$\tilde{s}(\omega, u) = \begin{cases} K_o\tilde{S}(\omega - \omega_o,\, u) + \bar{K}_o\,\bar{\tilde{S}}(-\omega - \omega_o,\, -u), & \text{range offset,} \\ K_o\tilde{S}(\omega,\, u - u_o) + \bar{K}_o(\bar{\tilde{S}}-\omega,\, -u - u_o), & \text{azimuth offset,} \end{cases}$$

(3.13a)

where

$$\tilde{S}(\omega, u) = \tilde{F}_o(\omega)e^{-i\omega\tau_o}\sum B_n(u\,;\, x_o)\exp\left[\frac{i}{2k}(u - nu_s)^2 + i(u - nu_s)x_o\right].$$

(3.13b)

The offset frequencies are chosen in the following way. (Reference to Fig. 3.2 will be helpful.) Consider, first, range offset operation. If the bandwidth of F is Δ, then to avoid overlapping of spectra and the attendant clutter, we want $\omega_o \geq \frac{1}{2}\Delta$. Recorder bandwidth limitations generally force us to the lower bound. In azimuth offset operation $\Omega \equiv kX$ is the bandwidth of (largest) adequately sampled segment of B; then for the analogous reason we want $u_o \geq \frac{1}{2}\Omega$ and are forced to the lower bound.

The sampling angular frequency is different in the two cases. For range offset operation we need

$$f_s(\text{ro}) = \Omega \quad \left(\frac{\text{rad}}{\text{sec}}\right)$$

by the definition of Ω, in order that the adequately sampled pieces of the azimuth alias spectra do not overlap. In azimuth offset operation we must have

$$f_s(\text{ao}) = 2\Omega \quad \left(\frac{\text{rad}}{\text{sec}}\right)$$

† We prove Eq. (3.13a) for the range offset mode; the argument is similar for the azimuth offset. We have

$$\begin{aligned} \tilde{S}(\omega, u) &= \mathscr{F}\{s(t, x)\} \\ &= \tfrac{1}{2}\mathscr{F}\{K_o S(t, x)e^{-i\varpi_o t} + \bar{K}_o\bar{S}(t, x)e^{-i\varpi_o t}\} \\ &= \frac{K_o}{2}\tilde{S}(\omega - \omega_o, u) + \frac{\bar{K}_o}{2}\bar{\tilde{S}}(-\omega - \omega_o, -u). \end{aligned}$$

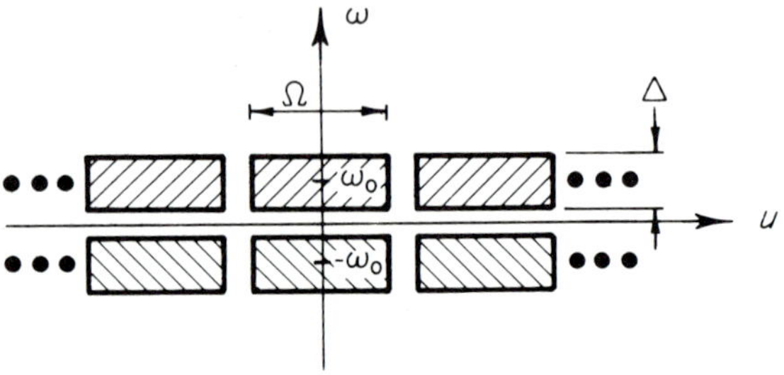

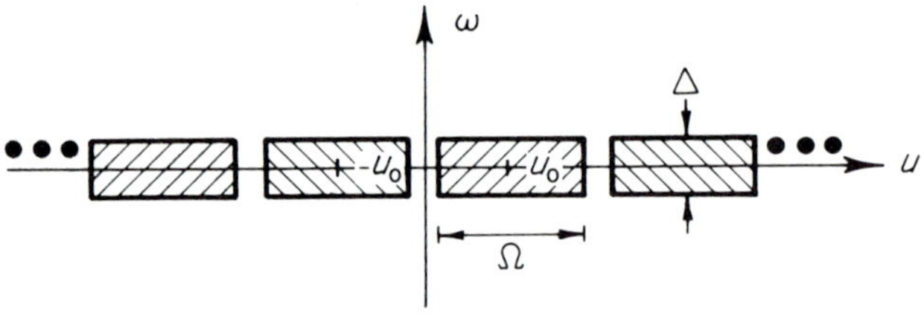

FIG. 3.2 The two-dimensional spectra of the received signal. (Only the adequately sampled piece of the aliases are shown, to avoid confusion.) Cross-hatch right to left: contributed by first term of $\tilde{S}$; cross-hatch left to right: contributed by second term of $\tilde{S}$. (a) two-dimensional spectrum in range offset case; (b) two-dimensional spectrum in azimuth offset case.

in order to make "holes" of width Ω in the spectra of the sampled RF signal so that, after coherent mixing, the "positive" and "negative" spectra can interleave without overlapping.

3.4. MATCHED FILTERING: POINT SCATTERER

If the two-dimensional signals due to a point scatterer were received along with an additive two-dimensional Gaussian noise process, with white spectral density over the frequencies of interest, then the optimum, or matched complex filter would be $\bar{\tilde{S}}(\omega, u)$ with $\tau_o \equiv x_o \equiv 0$ on the appropriate offset. However, the noise present is actually much different: it was derived from a one-dimensional noise process, assumed white. It can be shown† that in this case the filter reduces to

$$\bar{F}_o(\omega)B_o(u\,;0)\exp\left(\frac{-iu^2}{2k}\right)$$

on the appropriate offset frequency.

† See App. 4A.

The squared magnitude of the complex output, i.e., the squared envelope of the optimally processed real signal, is given by

$$|q(\tau, \mu)|^2 = |K_o|^2 \left| \int e^{i\omega\tau} |\tilde{F}_o(\omega)|^2 \, d\omega \right|^2$$
$$\cdot \left| \int e^{iu\mu} B_o(u:0) \exp\left(\frac{-iu^2}{2k}\right) \Sigma B_n(u\,;\mu) \right.$$
$$\left. \cdot \exp\left(\frac{i}{2k}(u - nu_s)^2 - inu_s\mu\right) du \right|^2 \tag{3.14}$$

for range offset operation, where

$$\tau \equiv t - \tau_o, \qquad \text{and} \qquad \mu \equiv x - x_o.$$

The function $|q|^2$ is the ambiguity function of the SAR radar when operated as described, with a range offset. The azimuth linear FM rate k is a function of delay; except for this, the ambiguity function factors with the two complex modulations determining the behavior in the τ and μ dimensions, respectively.

Recalling that B has infinite extent, $B_o(u; 0)$ passes only Ω-width segment centered about zero. Each term is matched in phase (but, except for $n = 0$, not in amplitude) and hence will essentially compress to a pulse-like shape at an azimuth location determined by the linear phase delay:

$$\mu_n = n\frac{u_s}{k} = nX. \tag{3.15}$$

These are the locations of the azimuth ambiguities, $n \neq 0$.

Of special interest is the response of $|q|^2$ about zero:

$$|q(\tau, \mu)|^2 \approx |K_o|^2 \left| \int e^{i\tau\omega} |\tilde{F}_o(\omega)|^2 \, d\omega \right|^2 \cdot \left| \int e^{iu\mu} |B_o(u\,;0)|^2 \, du \right|^2. \tag{3.16}$$

It will be sufficient for present purposes to note the peak response of each ambiguity—assumed to be at approximately μ_n:

$$|q(0, \mu_n)|^2 \approx |K_o|^2 \, \|F_o\|^4 \left| \int B_o(u\,;0) B_n(u;\mu_n) \, du \right|^2. \tag{3.17}$$

For azimuth offset operation the SAR ambiguity function is somewhat more complicated because both the positive and negative u spectra can be passed by the complex filter. We find the squared magnitude, or squared envelope, of the output signal to be

$$
\begin{aligned}
|q(\tau, \mu)|^2 = \Bigg| K_o \int e^{i\omega\tau} |\tilde{F}_o(\omega)|^2 \, d\omega \int e^{i\mu u} B_o(u; 0) \exp -i\left(\frac{u^2}{2k}\right) \\
\cdot \Sigma B_n(u; \mu) \exp\left(\frac{i}{2k}(u - nu_s)^2 - inu_s\mu\right) du \\
+ K_o \int e^{i\omega\tau}[F_o(\omega)]^2 \, d\omega \int e^{i\mu u} B_o(u; 0) \\
\cdot \exp -i\left(\frac{u^2}{2k}\right) \cdot \Sigma B_n(-u - 2u_o; \mu) \\
\cdot \exp\left(\frac{-i}{2k}(-u - 2u_o - nu_s)^2 - i(nu_s + 2u_o)\mu\right) du \Bigg|^2
\end{aligned}
\tag{3.18}
$$

where we remind ourselves that now $f_s = 2\Omega$.

It can be seen that the first integral has a behavior much like that for the range offset case: it gives compressed pulse-like outputs but at locations

$$
\mu_n = n\frac{u_s}{k} = 2nX. \tag{3.19}
$$

The second integral is different in its behavior: each term is completely mismatched in phase in the sense that the filter is the conjugate of the correct one. This results in even greater dispersion in both dimensions about locations

$$
\frac{2u_o + nu_s}{k} = (2n + 1)X. \tag{3.20}
$$

We shall drop this line of discussion at this point, resuming it in Chapter VIII when we consider the problem of optimum design of the SAR system with respect to ambiguities. Until then we will neglect these ambiguities.

3.5. LINEAR CHARACTERIZATION OF SAR

Equation (3.10c) of Sec. 3.2 gives the two-dimensional recording of the signal received by the SAR system when operated with a periodic modulation at a rate that approximately adequately samples the azimuth modulation. We have considered the form of this signal, its spectrum (Sec. 3.3), and its matched filtering (Sec. 3.4) for a point-scattering object.

It is important to note that any two-dimensional linear, quasi-invariant

("LQIV") processing commutes with the integral of Eq. (3.10c), giving for the SAR system image $|I|^2$, where

$$I(t, x) = \mathcal{C} \int dx' \int d\zeta \; q(t - \zeta, x' - x; \zeta) g(x, \zeta; \ldots). \tag{3.21}$$

In Eq. (3.21) $q(t, x)$ is the SAR system impulse response, or ambiguity function, ignoring ambiguities. There appears to be no reason not to choose the LQIV processor so that $q(t, x) = q_1(t)q_2(x; t)$. In particular, we saw in Sec. 3.4 that, if the processor is the matched filter, then q is so factored with $q_1 = F_o(t) * \bar{F}_o(-t)$ and $q_2(x; t) = B_X(x; t) * \bar{B}_X(-x; t)$.

We now consider in a simplified way the linear transfer function that characterizes practical SAR systems operating within their linear range and with a periodic modulation. We shall assume for concreteness that the data is recorded in the conventional two-dimensional format, and that it is processed by a coherent optical system.

In Sec. 3.4 it was shown that the output of the matched filter processor is, for a point scatterer's input signal,

$$|q(\tau, \mu)|^2 = |K_o|^2 \left| \int e^{i\omega\tau} |\tilde{F}_o(\omega)|^2 \, d\omega \right|^2 \cdot \left| \int e^{iu\mu} |B_o(u)|^2 \, du \right|^2,$$

where for simplicity we have ignored the ambiguities. We see that the output is a product of a function only of delay τ and a function only of azimuth position μ.† For this reason, we can consider these dimensions separately; it is sometimes convenient to describe the function of τ as the "output of the range *channel*" and similarly for azimuth.

It is possible to describe practical systems in this manner as long as all subsystems impulse responses can be factored. For example, the recorder may be described by a two-dimensional impulse response which is a product, e.g., $\exp(-\alpha r^2) = \exp(-\alpha\tau^2)\exp(-\alpha\mu^2)$.

Range channel. The range channel can be described by the block diagram of Fig. 3.3. The input signal is a point scatterer, modeled as a "delta function"; its convolution with the transmitted signal gives the physically returned signal. The radar receiver's transfer function is the next block.

The recorder is taken as a CRT, relay lens, film system. The CRT may be characterized by a Gaussian impulse response determined by the phosphor spot size under appropriate operating conditions. A lens can be characterized by a low pass impulse response which is easily measured. Films are also

† It is true, as we have said many times, that the linear FM rate k of the azimuth signal depends upon τ—but in such a weak way that it can be ignored when discussing the impulse response near the center of the ambiguity plane.

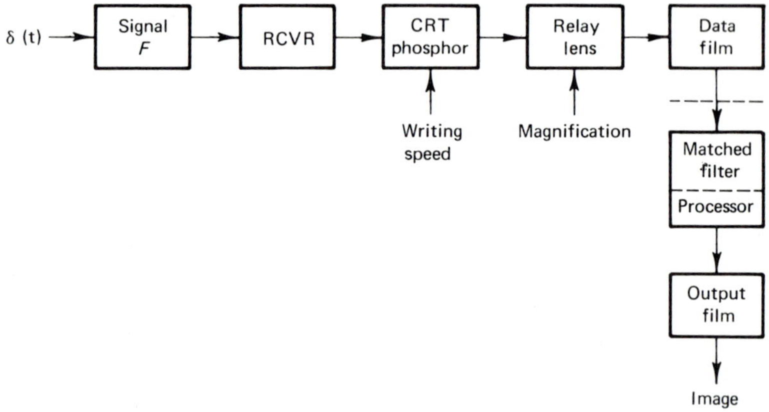

FIG. 3.3 The linear characterization of the range channel of a typical synthetic aperture radar system.

characterized by low pass transfer functions, under controlled exposure and development conditions. The magnification, if any, of the relay lens must be taken into account, as must the writing, or sweep, speed of the CRT "spot."

In order to account for the CRT sweep speed μ and relay lens magnification M parameters, we examine the recording operation. The range channel signal at the input to the recorder—e.g., at the control grid of the CRT—we denote $S_1(t)$. Any recorder has a spatial impulse response† $h(x)$—e.g., the phosphor spot of the CRT. Thus, the signal available for recording —e.g., at the face of the CRT—is $S_2(t, x) = S_1(t)h(x)$. The motion of the CRT spot scanning is described by $y = x + vt$, where y is the linear dimension on the recording medium—e.g., ordinary photographic film. An incremental contribution to the record at any point y is

$$dS_{r1}(y) = S_1(t)h(x)\,dt, \qquad x = y - vt;$$

summing over all time we have

$$S_{r1}(y) = \int dt\; S_1(t)h(y - vt),$$

or, in spatial variables

$$S_r(y) = \int dz\; S(z)h(y - z), \tag{3.22}$$

where

$$S_r(y) \equiv vS_{r1}(y)$$

and

$$S(z) \equiv S_1(s/v).$$

† If it has a significant temporal impulse response this can be included as a transfer function after the receiver.

This result [Eq. (3.22)] is intuitively clear: the recorded signal is simply the convolution of the recorder impulse response in the range dimension with the input signal, appropriately scaled by the recording velocity.

The effect of the magnification M is easy to see: the signal $S_r(y)$ is not actually recorded but is imaged onto the recording medium. The image is obviously $S_r(y/M)$.

In order to compute the image of any input signal, we must know the transfer function of every block shown in Fig. 3.4. Then the overall transfer function is, of course, the product of all of the individual transfer functions. The impulse response, which is the inverse Fourier transform of the transfer function, is usually best computed by a digital computer. (In fact, some of the individual transfer functions may best be found by computer calculation—those for lenses particularly.)

For illustrative purposes and as a good approximation in practice, we can approximate each individual impulse response by a Gaussian function, of zero mean—corresponding to a low pass transfer function—and a variance. Then it is well known that the overall transfer function will be Gaussian, and its variance will be the sum of all the individual variances. The variance (actually its square root) is a good measure of the resolution capability of the range channel, and, thus, we get a good estimate of this in a simple way. Of course, we do not learn anything about the actual sidelobe level, for example; a more exact calculation is needed.

To point out the essentials, suppose the receiver is of sufficient bandwidth so that it may be ignored. The variance of the impulse response of the system is therefore

$$\sigma_r^2 = (M_r \sigma_P)^2 + \sigma_L^2 + \sigma_{FD}^2 + \sigma_C^2 + \sigma_{FI}^2 \qquad (3.23)$$

where σ_P^2, σ_L^2, σ_{FD}^2, σ_C^2, σ_{FI}^2, are the variances of the impulse responses of the phosphor, relay lens, data film, optical processor, and image film, respectively; M_r is the relay lens magnification in the range dimension, of course. The variance of the signal is

$$\nu^2 = \left(\frac{vM_r}{\Delta}\right)^2$$

where Δ is the temporal bandwidth, in Hz, of the radar signal modulation. The impulse response of the entire system therefore has variance

$$\begin{aligned}\theta^2 &= \sigma_r^2 + \nu_r^2 \\ &= (vM_r)^2\left[\frac{1}{\Delta^2} + \frac{\sigma_P^2}{v^2} + \frac{\sigma_L^2 + \sigma_{FD}^2 + \sigma_C^2 + \sigma_{FI}^2}{v^2 M_r^2}\right].\end{aligned} \qquad (3.24)$$

The factor $(vM_r)^2$ in Eq. (3.24) merely scales the output.

Equation (3.24) is useful in approximately specifying the bandwidths and other parameters that must be met for a given resolution. We observe the following facts:

1. By making the recorder sweep velocity very large, the degradation of resolution may be made very small. This is intuitively reasonable: increasing the sweep increases the spatial extent of the recorded signal, thereby decreasing its spatial bandwidth: the transfer function of the range channel thus is effectively increased. In practice, v cannot be indefinitely increased because of one or more of the following limitations: recorders have limited linear sweep ranges, the whole slant range interval of interest must be recorded, the recording medium (e.g., film) has a restricted extent, and the apertures over which the optical components perform satisfactorily is restricted.
2. By choosing a large magnification, the transfer functions occurring after the CRT face may be made unimportant for the same reason. In practice the limitations that prevent indefinite increase of magnification are: the recording medium, and usable optical apertures have limited extent.

Azimuth channel. The azimuth channel is described by the block diagram of Figure 3.4. We make the same assumptions regarding recording methods

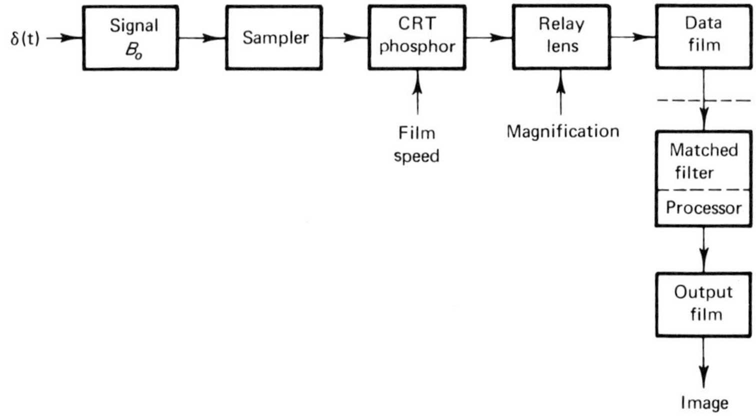

FIG. 3.4 The linear characterization of the azimuth channel of a typical synthetic aperture radar system.

as for the range channel. Because the azimuth signal is sampled, the channel differs slightly from the range channel. However, ignoring ambiguities, with a properly designed overall transfer function, we can omit the sampler from the block diagram. Also the azimuth signal bandwidth is extremely small relative to the signal bandwidth, so the radar receiver prior to recording may be assumed ideal.

The recording operation has the medium (e.g., film) moved at a velocity μ, analogously to the sweep velocity. It is possible to design astigmatic optics so that the azimuth dimension magnification M_a need not equal that in the range dimension, M_r, and so for other components. The impulse response of the processing system is generally

$$\sigma_a^2 = (M_a \sigma_P)^2 + \sigma_{La}^2 + \sigma_{FD}^2 + \sigma_{Ca}^2 + \sigma_{FI}^2, \tag{3.25}$$

and the entire system is

$$\theta_a^2 = (\mu M_a)^2 \left[\frac{1}{\Omega^2} + \frac{\sigma_P^2}{\mu^2} + \frac{\sigma_{La}^2 + \sigma_{FD}^2 + \sigma_{Ca}^2 + \sigma_{FI}^2}{\mu^2 M_a^2} \right].$$

Remarks concerning choice of μ and M_a in the range channel apply analogously here, except that instead of a recorder sweep extent limitation we have a possible limit on film speed due to mechanical problems, weight and/or volume restrictions, light intensity availability, and so on.

3.6. SIGNAL-TO-NOISE RATIO

We turn finally to the calculation of the signal-to-noise ratio, SNR. We assume that a matched filter processor is used and that the additive noise has a constant (N_0) spectral density over the spectral band of the received signal, the latter as given by Eq. (3.9). Further, we calculate the SNR for a "point" object—an object small with respect to the system resolution, but large relative to the wavelength, of reflectivity density g_o, and area α.

There is one detail not mentioned in Sec. 3.1 because it was unimportant until now: in order that the aperture illumination function not increase the energy of the generated signal we must require that

$$\|E_0\|^2 \equiv \iint_A |E_0(x, y)|^2 \, dx \, dy = 1. \tag{3.26}$$

This is accomplished in Eq. (3.9) if E_0 is replaced everywhere by $(E_0/\|E_0\|)$. Thus, with these substitutions we have for the returned complex signal

$$S(t) = \left(\frac{ik}{4\pi R_o} \right)^2 g_o \alpha F(t) B_v(0) B_h(vt; 0) \tag{3.27}$$

for a representative scatterer location at the center of the vertical beam.

As we shall see in Chapter IV, the SNR at the output of the matched filter is simply E_s/N_o where E_s is the energy of the (real) signal s:

$$\mathrm{SNR} = \frac{\frac{1}{2}\int |S(t)|^2\,dt}{N_o}.$$

If we assume, as is usually true in practice, that the signal modulation F has period T_s and that the azimuth modulation is approximately adequately sampled,† we can write

$$\int |F(t)|^2\,|B_h(vt)|^2\,dt = \frac{1}{T_s}\int |F_0(t)|^2\,dt \cdot \int |B_h(vt)|^2\,dt;$$

this is proved in Chapter IX (Property 9.4). Thus,

$$\mathrm{SNR} = \frac{1}{N_o}\left(\frac{ik}{4\pi R_o}\right)^4 (g_o\alpha)^2\,\frac{\int |F_0(t)|^2\,dt}{2T_s}\,|B_v(0)|^2\,\frac{\int |B_h(x)|^2\,dx}{v}. \tag{3.28}$$

We note that $\int |F|^2/2T_s$ is the average transmitted power, denoted P_{ave}.

This is not a very familiar looking form, and we proceed to simplify it. We first acknowledge the usual definition of scattering cross section, σ^2, for an object in the far field of the radar set [Kerr (1951), p. 455]

$$\sigma^2 \equiv 4\pi R_o{}^2 \left|\frac{E_s}{E_f}\right|^2.$$

Substituting directly for these fields as we derived them in Sec. 3.2, we find

$$\sigma^2 = \frac{\pi}{\lambda^2}\,|g_o\alpha|^2. \tag{3.29}$$

Keeping in mind the normalization, Eq. (3.26), we have by the definitions introduced in Sec. 3.2,

$$|B_v(0)|^2 = \left|\frac{\tilde{E}_{0v}(0)}{\|E_{0v}\|}\right|^4 = \left|\int_{-\Omega_v}^{+\Omega_v} \frac{E_{0v}(z)}{\|E_{0v}\|}\,dz\right|^4,$$

and

$$\int |B_h(x)|^2\,dx = \frac{R_0}{k}\int \left|\frac{\tilde{E}_{0h}(\xi)}{\|E_{0h}\|}\right|^4 d\xi.$$

Substituting these relations into Eq. (3.28), we find

$$\mathrm{SNR} = \frac{\sigma^2}{4\pi\lambda R_0{}^3{}_v}\cdot\frac{P_{\mathrm{ave}}}{N_o}\cdot\frac{\chi}{8\pi}, \tag{3.30}$$

† Inadequate sampling in a well-designed system has negligible effect on the SNR—see Chapter VIII—but, of course, does affect the signal-to-ambiguity ratio.

where

$$\chi \equiv \frac{\left|\int_{-\Omega_v}^{+\Omega_v} E_{0v}(z)\,dz\right|^4}{\|E_{0v}\|^4} \cdot \frac{\int |\tilde{E}_{0h}(x)|^4\,dx}{\|E_{0h}\|^4}. \tag{3.31}$$

We can reduce Eq. (3.30) to a still more convenient and familiar form, by examining χ as defined by Eq. (3.31). We can write

$$\frac{\chi}{8\pi} = \frac{A^2}{D_h} \cdot Q$$

where

$$Q \equiv \frac{1}{64\pi} \cdot \frac{\left|\int_{-1}^{+1} E_{0v}(y\Omega_v)\,dy\right|^4}{\left(\int_{-1}^{+1} |E_{0v}(y\Omega_v)|^2\,dy\right)^2} \cdot \frac{\int \left|\int_{-1}^{+1} e^{-iu\zeta} E_{0h}(\zeta\Omega_h)\,d\zeta\right|^4 du}{\left(\int_{-1}^{+1} |E_{0h}(\zeta\Omega_h)|^2\,d\zeta\right)^2}$$

is a number dependent only upon the shape of the antenna illumination functions. E.g., if the illumination functions are uniform, it is easy to see that

$$Q = \frac{1}{4\pi} \int \left(\frac{\sin u}{u}\right)^4 du = \frac{1}{6}. \tag{3.32}$$

We now have our final form

$$\boxed{\mathrm{SNR} = \frac{A^2\sigma^2}{4\pi\lambda R_o^{\,3}} \cdot \frac{P_{\mathrm{ave}}}{N_o} \cdot \frac{Q}{vD_h}} \tag{3.33}$$

for the signal-to-noise ratio of the SAR system at the center of the vertical beam: it decreases in an obvious way away from the vertical beam center. Generally, Q will be of the order of unity, roughly; for a uniformly illuminated aperture we saw above that $Q = \frac{1}{6}$.

A simple approximate derivation of SNR. We now derive in a rather simple way an approximate relation for the received power and signal-to-noise ratio for synthetic aperture radars. Because of the obvious importance of this relation, it is worth our time. We begin with the "classical" radar range equation, and, further, we will treat the azimuth compression operation as an addition of power. The radar range equation gives for the returned instantaneous signal power of a single pulse

$$P_{r1} = \frac{P_{t1} A_e^{\,2} \sigma^2}{4\pi\lambda^2 R^4},$$

where $P_{t1} = \frac{1}{2}|F(t)|^2$ is the transmitted power of the pulse. We approximate the number of pulses added by the product of the sampling rate, f_s pps, times the nominal beamwidth (in seconds):

$$N = f_s \frac{R\,(\lambda/D_h)}{v}.$$

If the antenna gain had equal amplitude over its beamwidth then the total received instantaneous power is

$$P_r = \frac{A_e^{\,2}\sigma^2}{4\pi\lambda R^3 v D_h} \cdot f_s \cdot \tfrac{1}{2}|F(t)|^2.$$

If we now match filter process the signal, the final signal-to-noise ratio is the ratio of signal energy to the amplitude N_o of the white noise spectral density:

$$\text{SNR} = \frac{A_e^{\,2}\sigma^2}{4\pi\lambda R^3 v D_h} \cdot \frac{P_{\text{ave}}}{N_o},$$

where again we have set $P_{\text{ave}} \equiv f_s \cdot \frac{1}{2}\|F\|^2$.

Interpretation. In using this equation to determine parameter dependencies of the SNR, some discretion is required. For example, we pointed out that the scatterer is replaced by an equivalent isotropic scatterer with scattering cross section σ in accordance with convention. Generally, σ depends upon λ—even, sometimes, when g_o does not. Further, σ generally depends upon angular orientation, polarization, and so forth, as we remarked.

For a point scatterer it is clear why $\text{SNR} \approx D_h^{-1}$. For as D_h increases, the beamwidth λ/D_h decreases, and the number of pulses added decreases as D_h^{-1}, all other parameters fixed. It is also clear why SNR is proportional to R_o^{-3} instead of the usual R_o^{-4}: the number of pulses added from a scatterer is proportional to R_o.

Again, for a point scatterer, the apparent dependence on v is less clear. But we recall $P_{\text{ave}} \propto f_s \propto v$: thus, if we hold the energy per transmitted pulse constant, we see that SNR is independent of v.

If, in fact, the energy per pulse, E_F, is a restraint—usually P_{ave} is the restraint—then the SNR is better written in the form

$$\text{SNR} = \frac{\delta D_v^{\,2}\sigma^2}{2\pi\lambda R^3} \cdot \frac{E_F}{N_o}$$

where $\delta = 1$ if a range offset is used, and $\delta = 2$ if an azimuth offset is used.

Extended terrain. When computing the SNR for extended terrain an approximation is used that the scattering cross section σ is the scattering cross-section density σ_o—a generally wavelength dependent, tabulated parameter—times the area of a resolution cell. Thus,

$$\sigma^2 \approx \sigma_o^2 \rho_\tau \frac{D_h}{2},$$

where ρ_τ, the delay resolution, is determined essentially by the bandwidth of the radar's physical signal modulation. If P_{ave} is the significant restraint we have

$$\text{SNR} = \frac{A_e^2 \sigma_o^2 \rho_\tau}{8\pi\lambda R^3 v} \cdot \frac{P_{\text{ave}}}{N_o}.$$

The dependence on D_h is now only via the aperture area.

APPENDIX 3A: REFLECTIVITY DENSITY CALCULATION BY THE METHOD OF PHYSICAL OPTICS

We show now that modeling the scattering mechanism with a "reflectivity density" (Sec. 3.2) is quite good for objects with good conducting surfaces that are large and smooth relative to the radiated wavelength. Using the method of physical optics, we shall explicitly calculate the reflectivity density.

The method of physical optics represents, for amenable objects, a very great improvement over geometrical optics, without a great deal of mathematical difficulty: this method accounts for wavelength dependence in its usual formulation and can be generalized to account for polarization [Mentzer (1955), Sec. 2.3.2]. The method applies to objects of major dimensions and radii of curvature somewhat greater than a wavelength—just as does the method of geometrical optics—but the object must be a good conductor: this appears to be the restriction we must accept in order to get more properties of the scattered wave without great effort at computation.

It is not difficult to represent the scattered field as an integral over the surfaces of scattering objects. The integrand involves the electric and magnetic fields at the object boundary—which are of course generally unknown. When the surface is of infinite conductivity the integrand simplifies to a form requiring a knowledge—or estimate—of the surface currents on the scattering object. For small objects the interaction of these currents makes their calculation or estimation very difficult. But for objects large relative to the wavelength, the total tangential component of the magnetic field is approximately twice the tangential component of the incident field alone.†

† From Maxwell's equation one can show that the surface current density $\mathbf{J} = \mathbf{n} \times \mathbf{H}$, where $\mathbf{n}$ is the surface normal, and $\mathbf{H}$ the magnetic field.

The scattered magnetic field vector $\mathbf{H}_s$ can be expressed by the vector form of the Kirchoff-Huygens diffraction integral already discussed in scalar form:

$$\mathbf{H}_s = \frac{-ik}{4\pi R} \iint_S \mathbf{n}_o \times (\mathbf{n} \times \mathbf{H}_t) e^{-ikr}\, dS. \tag{3A.1}$$

In this relation, $\mathbf{H}_t$ is the *total* (necessarily tangential) magnetic field on the surface of the scattering object whose surface unit normal is $\mathbf{n}$ and boundary is S; $\mathbf{n}_o$ is the direction of propagation of the incident field. We have assumed that the distance to the object from the source/receiver is large relative to the size of the object [Kerr (1951), Sec. 6.2]; the scattered field then can have only a negligible longitudinal field.

If the incident field is $\mathbf{H}_i \equiv H_i \mathbf{a}$, then the method of physical optics is to take

$$\begin{aligned} \mathbf{H}_t &= 2\mathbf{H}_{it} \\ &= 2H_i[\mathbf{a} - \mathbf{n}(\mathbf{a} \cdot \mathbf{n})]. \end{aligned}$$

For narrow beam SNR systems we may assume that $\mathbf{a} \cdot \mathbf{n}_o \approx 0$: that is, the radiated field has only a transverse polarization of any importance, and it is approximately orthogonal to any illuminated object. Then it is easy to show by vector identities [Kerr (1951), p. 463] that:

$$\mathbf{n}_o \times (\mathbf{n} \times \mathbf{a}) \approx -(\mathbf{n}_o \cdot \mathbf{n})\mathbf{a}$$

and hence that

$$\mathbf{n}_o \times (\mathbf{n} \times \mathbf{H}_t) \approx -2H_i(\mathbf{n}_o \cdot \mathbf{n})\mathbf{a}. \tag{3A.2}$$

Thus finally

$$\mathbf{H}_s \approx \mathbf{a} \frac{ik}{4\pi R} \iint_S e^{-ikr} H_i \cdot 2(\mathbf{n}_o \cdot \mathbf{n})\, dS. \tag{3A.3}$$

Recalling Eq. (3.5), we see that *the method of physical optics is equivalent to taking* $2(\mathbf{n}_o \cdot \mathbf{n})$ *as the reflectivity density over the surface S of the object.* The remaining steps to find the SAR system's recorded signal for such an object are analogous to those performed in Chapter III to derive Eq. (3.10c) from Eq. (3.5). Thus, if the radiated SAR modulation is a periodic repetition of the modulation F_o, and the azimuth modulation is approximately adequately sampled, we have as the SAR system received data

$$S(t, x) = \left(\frac{k}{4\pi R_o}\right)^2 \iint_S dS' F_o\left(t - \frac{2y'}{v}\right) \cdot B_v(z') B_h(x' - x; y') N(x', y', z'), \tag{3A.4}$$

where in turn

$$N(x, y, z) \equiv 2(\mathbf{n}_o \cdot \mathbf{n})(x, y, z), \qquad (x, y, z) \in S. \tag{3A.5}$$

Matched filtering. If the received data, Eq. (3A.4), is processed by the linear, quasi-invariant filter matched to a point object (Sec. 3.4), then we have, for the SAR system image, $|I(t, x)|^2$, where

$$I(t, x) = \left(\frac{k}{4\pi R_o}\right)^2 \iint_S dS' q_1\left(t - \frac{2y'}{v}\right) q_2(x - x'; y') N(x', y', z'). \quad \text{(3A.6)}$$

Here we have written the system impulse response, i.e., ambiguity function, in the quasi-factored form that results when a periodic modulation is used (Secs. 3.4 and 3.5).

The result (Eq. 3A.6), is an interesting and important result, for it gives us quite exactly the SAR system image of an object in this class. There are no other precise results available. It characterizes the effect on the image of the geometry of the surface of an object whose electrical properties are constant. For example, *it is possible for a perfectly conducting object to scatter energy that is in fact received by the SAR system, and yet there will appear no image whatsoever.*

We note once again that Eq. (3A.6) is a specific manifestation that, because of the assumed linearity of all mechanisms involved—including that of scattering, *only the overall system impulse response $q_1 q_2$ determines the image for a given object*: it does not matter whether or not dispersed pulses were radiated or whether or not azimuth resolution was obtained by "brute force" or by the SAR technique.

Example. Suppose that the object is a corrugated sheet whose equation for its surface S is given by

$$y - R_o - y_o \sin u_1 x = 0, \qquad -\tfrac{1}{2}z_o \leq z \leq \tfrac{1}{2}z_o,$$

and S does not extend beyond z_o, which is of the general order of the system resolution—it does not interest us exactly what it is. Suppose also that y_o is small relative to the slant range resolution of the system, so that

$$q_1\left(t - \frac{2y'}{v}\right) \approx q_1\left(t - \frac{2R_o}{v}\right), \qquad y' \in S.$$

Then it is straightforward to calculate

$$2(\mathbf{n}_o \cdot \mathbf{n}) = 2[(u_1 y_o \cos u_1 x)^2 + 1]^{-1/2} \equiv N_1(x) \quad \text{(3A.7)}$$

and

$$S(t, x) = \left(\frac{k}{4\pi R_o}\right)^2 q_1\left(t - \frac{2R_o}{v}\right) \int dx' q_2(x - x'; R_o) N_1(x'). \quad \text{(3A.8)}$$

The integral extends over the x-length of the sheet, of course.

One important thing to note about Eq (3A.8) is that $N_1(x)$ can be mainly composed of a fluctuating part with fundamental spatial frequency u_1. The spatial frequency u_1 can be large with respect to the spatial frequency pass-band of the SAR system, in which case the *sheet may not be seen.*

Conclusions. We have derived in this Appendix the SAR image, Eq. (3A.6), for objects whose scattering mechanisms can be described by the method of physical optics: objects of size and radii of curvature at least somewhat larger than the wavelength, and of good conductivity. We have shown that such objects are modeled by a *real* reflectivity distribution over the surface of the object. We also note that wavelength variation over the region of wavelengths small with respect to object size and radii of curvature will not change the nature of the image. Also variation of polarization has no effect on the image.

COMMENTS AND REFERENCES

The received signal. We have used the Kirchoff-Huygens principle: for a discussion of it see, e.g., Wolf and Born (1965). For a discussion of the scattering mechanism see, e.g., Kerr (1951). A simplified derivation for a point scatterer was given by Harger (1965b).

The signal spectrum. The clarification of the role of the offset frequencies for two dimensional complex signals was first explained, at least in the present context, by E. N. Leith and C. J. Palermo in unpublished work; later these ideas contributed to the remarkable invention of practical holography: Leith and Upatnieks (1962, 1965).

Matched filter. The discussion follows Harger (1965b).

Signal-to-noise ratio. Various approximate forms of the SNR equation have been known for an indeterminant time. For the conventional definition of scattering cross section see, e.g., Kerr (1951).

Reflectivity density calculation. Our discussion of the method of physical optics follows Mentzer (1955).

IV

System Design A

4.1. SAR SYSTEM MODELS

In this chapter we wish to discuss rational approaches to designing the SAR system. We shall begin by considering sensible procedures without taking into account any error, or noise, sources of the system, then we will discuss criteria that account for *additive* errors, especially random processes. We shall defer until later (Chapters VIII and X) discussion of optimal procedures in the face of other types of errors.

When system noise is not considered, criteria based on resolution—defined in a variety of ways—are widely used in communication work. We shall discuss two representative resolution criteria that are sufficiently illustrative. A widely used mathematical criterion is least-square error; this is also discussed.

When additive system noise is accounted for, a popular communication criterion is maximum signal-to-noise ratio (SNR_{max}). Next, the extension of the least-square error method to random signals and noise is discussed. Last, the very interesting method of maximum likelihood is described; unlike all the previous criteria, we do not assume a linear system to begin with, but end up with one anyway (assuming *Gaussian* additive noise, however).

Finally, we compare these results and note that they agree sufficiently to provide a fairly firm basis for system design.

In order to discuss optimum system design we, of course, need an analytic model for the SAR system; such a model was partially derived in Sec. 3.2. We assume that the received signal is available for processing in the two-dimensional form given by, e.g., Eqs. (3.10) and (3.11). Recall that a periodic signal modulation was assumed in order to obtain these forms. We ignore the presence of aliases created by the sampling operation: their presence will be accounted for in Chapter VIII. Of course, aliases are not a problem of importance in some SAR systems. As we remarked (Sec. 3.2) the received signal is a transformation, or mapping, of the reflectivity density that is linear and quasi-invariant.

To complete the SAR system model in the noise/error free case, we need to add a processor: there appears to be no reason not to assume that it also is a linear, quasi-invariant—"LQIV"—operation. We denote it by an impulse response $H(t, x; t)$, the third argument representing the relatively slow change in delay coordinate that will be necessary because of the slow change with delay coordinate of the azimuth modulation $B_h(x; t)$. The transfer function of the system will be denoted therefore $\tilde{H}(\omega, u; t)$, the two-dimensional Fourier transform operating on just the first two arguments of H; this corresponds to the spectrum of the received signal $\tilde{F}(\omega)\,\tilde{B}_X(u; t)$. We shall sometimes suppress this slow t-variation in the notation. This noise-free model is shown in Fig. 4.1.

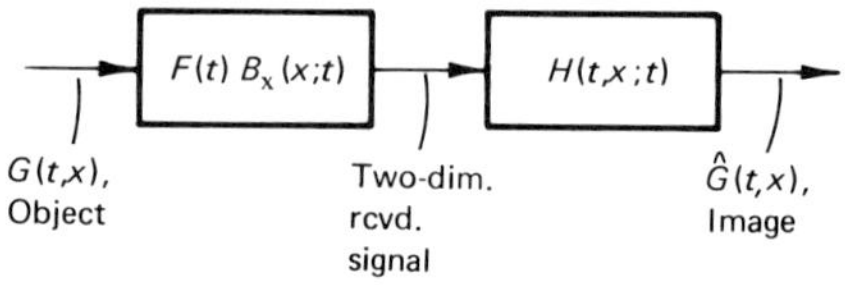

FIG. 4.1 Noise and error free model of SAR system: F is the complex signal modulation; $B_X(x; t)$ is the complex azimuth modulation, and $H(t, x; t)$ is the linear, quasi-invariant processor.

A more reasonable and practically significant SAR system model includes the additive noise that is unavoidably present in such systems. In microwave systems there is always thermal noise; since the characterization of this noise is well known in radar, we do not discuss it except to recall that it is a Gaussian random process, and that the effective noise temperature of an SAR system is usually determined by the temperature of the terrain. However, we do need to find out what the nature of this noise is when recorded in a two-dimensional record: it is not hard to show (see App. 4A) that this "RF noise, n_1," as we shall call it, is a two-dimensional Gaussian random field whose spectral density

$$P_1(\omega, u) = \begin{cases} \dfrac{u_s}{2\pi} N_o, & |u| < \frac{1}{2}u_s, \\ 0 \quad , & |u| > \frac{1}{2}u_s, \quad \text{all } \omega. \end{cases}$$

In particular, we note that the covariance distance of such noise will be very small relative to the SAR system aperture (T_s, X): this fact simplifies some of the forms of the optimum systems.

In practice, the data-storage operation introduces noise that can be modeled approximately by additive noise. For example, the grain noise of ordinary photographic film can be so modeled sufficiently well for our purposes. All additive noise sources other than the RF noise n_1 we shall denote n_2 and assume the latter are given in two-dimensional form; the sum of n_1 and n_2 we denote n, the additive noise. Often, it is reasonable to assume that n, like n_1, has a spectral density constant over the frequencies of interest (see Fig. 4.2).

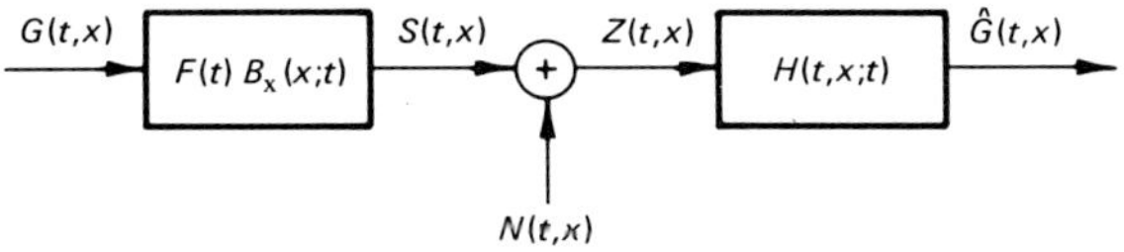

FIG. 4.2 Model of SAR system including additive noise.

Sometimes it will be informative and reasonable to assume that the terrain reflectivity density is composed of two additive parts: $g \equiv g_1 + g_c$. g_1 could represent a specific object of interest, and the remainder of the terrain would be g_c: the latter is often referred to as "clutter." In order to design a system to suppress the clutter's degradation, we must be able to characterize it in some way. There appears to be insufficient knowledge to do this. We shall here consider it a random process with a known (or to be measured) spectral density (Sec. 4.3.2).

The systems optimum under several criteria that account for additive noise will be found to be quite similar and therefore provide a sensible base for SAR system design. Details of this design will be the subject of Chapters V and VIII.

It is important to observe at the outset, that very often the purpose of the SAR system is not to form an image of the entire terrain, but only special objects of interest. Specifying in a useful way what these objects are is a major initial difficulty in the system design. In fact a common use of the SAR system would be to *detect* objects of interest and *estimate* their position: we shall see that, in such cases, the optimum system forms an image only in preserving (approximately) the geometrical arrangement of these special objects.

Since we are assuming the azimuth modulation has been approximately adequately sampled, we will be talking about bandpass signals and linear systems. It is natural to use the complex representation for such signals and systems. The reader unfamiliar with this representation may proceed, ignoring it, and still get the essential content of the discussion, and then, if he likes, see, e.g., Brown (1963b), or Berkowitz (1965). For example, if f is a real signal (system), then F is the complex representation of f:

$$f = \mathrm{Re}\{Fe^{i\omega_c t}\}.$$

4.2. CRITERIA FOR NO NOISE

When error sources in a system can be ignored, the customary statistical criteria are inapplicable. But there are still logical questions about system design, and rational design procedures are necessary. In the SAR system, the basic question is: what are we looking for, and how extensively are we willing to commit the system to finding it? Characterizing "what we are looking for" in a useful way is very difficult; even if we have a specific object in mind, it is a difficult matter to figure out how this object appears to the SAR system. This problem is often avoided by designing the system to do "well," in some sense, when the scattering object is a "point": that is, one attempts to shape the system impulse response. All such procedures we shall term "resolution criteria."

4.2.1. Resolution Criteria

Resolution criteria are criteria based on desired properties of the system impulse response, usually defined for ideal, or noise-free operation, and for LIV systems. The basic idea is simply that one supposes that, the more closely the system impulse response resembles a delta function, the more closely the system's output/image should resemble the system's input/object. The major defect of the criterion is that it is not usually clear precisely how properties of the system impulse response affect errors of various kinds in the image. The major advantage of the method is it represents a rather minimal "commitment" on a data processing criterion and, in that sense, allows perhaps greater flexibility in subsequent use of the data.

Other advantages of the criterion are that: (1) it allows simple formulation and solutions; (2) it presupposes a linear processing structure for which there are well established physical synthesis procedures; (3) the output of the processor is an image; (4) it can be generalized to some extent to noisy situations; (5) it gives us a logical procedure in the noise-free case when statistical criteria are useless. Other disadvantages of the criterion are that when noise of some type is present, as it always is, (1) the criterion does not a priori agree with the common statistical optimality criteria, and (2) the restriction to linear processors precludes optimality for some types of noisy situations. Despite these disadvantages, we are not surprised that the processor structure generally is similar to those found by various criteria for various kinds of weak noise.

There is another feature of resolution as a criterion: it is a "local" criterion, in that we are trying to make the system perform well at each point. This is attractive when we are trying to identify the image of relatively small, isolated objects, as we often are. The least-squares criterion, to be discussed below, is a "global" criterion, as usually defined, in that it concerns itself with average errors, and we have no detailed controls over where the larger errors occur: it might happen that they turn out to be probable about objects of special interest!

Resolution properties. There are several characteristics of system impulse response that can be of interest: value of maximum response, location of maximum, width of major lobe, value and location of secondary maxima, energy in major lobe, etc. Generally, all of them are of interest when investigating the effects of various kinds of errors on system resolution. But for a system design criterion, the most common resolution criterion is based on two properties: width of major lobe, and maximum sidelobe level. An analytically convenient resolution criterion that is a little indirect is the property: percent energy in the desired resolution cell. There are other analytically tractable criteria but we shall not be concerned with them here.

A simple approach. A well-known general fact is that minimum main lobe width is inversely proportional to bandwidth. This can be quantified in various ways: e.g., if radius of gyration is used as a measure of the spread of the system impulse response and also its bandwidth, the product—the "time-bandwidth" product, TBP—is bounded below by $\frac{1}{2}$. The available bandwidth is set by practical restraints. The limit on the spectral width in the range dimension is determined by the "state-of-the-art" in generating and processing large bandwidth—usually large TBP signals. The spectral width in the azimuth dimension is restrained by any system requirement on unambiguous range interval, as we have seen in Eq. (1.1), and therefore may not be a free parameter.†

A very simple criterion—hardly deserving of the name— is to make the system transfer function constant over the available system bandwidth. The idea is simply that a delta function is presumed to be the ideal system impulse response and this is the "best" approximation to its constant spectrum that we can get.

Referring to Fig. 4.1 for the SAR system model, we see that the SAR processor transfer function

$$\tilde{H}(\omega, u; t) = \frac{1}{\tilde{F}(\omega)\tilde{B}_X(u; t)},$$

for (ω, u) such that $\tilde{F}(\omega)\tilde{B}_X(u; t) \neq 0$.

† Of course, the system azimuth bandwidth cannot indefinitely be increased because its upper band is set by vehicle velocity. The corresponding minimum resolution is about $\lambda/4$ (Sec. 2.5).

We note first that the conjugate phase, $-\text{Arg } B_X$, arises, providing the compression operation that we know should be accomplished. Next, we see that the far-field pattern of the antenna—and hence $|\tilde{B}_X|$—cannot be flat: this implies that this system will unfortunately emphasize the white additive noise that will unavoidably be present. Finally, the system impulse response will be (for a rectangular transfer function) a product of $(\sin \zeta/\zeta)$-type functions: this means that the maximum sidelobes will be relatively large (-13.4 db) and that a relatively large percentage of the impulse responses energy will be in the sidelobes (about 10%).

Resolution with sidelobe energy control. The sidelobe level that resulted from the "simple-minded" approach above is usually not regarded as satisfactory. For example, a point reflecting object would give a sidelobe that would tend to mask the image of an adjacent point object, if the latter's returned signal strength was of the order of the sidelobe level of the former object. It is, therefore, of interest to formulate a criterion that directly attempts to control sidelobe level. Unfortunately this property of the system impulse response is not amenable to useful mathematical characterization. Thus, one must choose criteria that indirectly control sidelobe levels. There are several ways to do this, but we shall consider just one; it leads to typical results and is the most elegant mathematically.

We note that if we could manage to contain the system impulse response entirely within the desired resolution cell, we would have achieved the ideal aim. (Note that the fluctuations of the impulse response within this cell are of no interest.) Of course, we know that a finite bandwidth function cannot be confined to such a finite cell; however we can formulate the criterion: maximize the "amount" of the impulse response in the cell, using "energy" as the measure of "amount." The energy measure is chosen both because it is reasonable, and because it is mathematically very fruitful.

If we assume the system impulse response to be in factored form, $q_1(t)q_2(x)$, then we really have two one-dimensional problems. For example, we want to choose $q_1(t)$, a band limited function, so as to have maximum percent energy in the resolution cell, I_s say. In mathematical language, our problem is

$$\max_{f_1} \frac{\int_{I_s} |q_1(t)|^2 \, dt}{\int |q_1(t)|^2 \, dt},$$

subject to

$$q_1(t) = \frac{1}{2\pi} \int_{-\Omega_s}^{\Omega_s} e^{i\omega t} f_1(\omega) \, d\omega.$$

The purpose of the latter equation is to force q_1 to be a function bandlimited to $2\Omega_s$, the signal bandwidth. The same holds for the azimuth dimension problem.

We shall have occasion to discuss this mathematical problem later, in connection with the ambiguity problem (Chapter VIII) and shall therefore simply give the solutions here. The problem must be solved numerically, and the solutions $\{f\}$ have already been named "prolate spheroidal wave functions (PSWF)." They are shown in Figs. 8.4a and 8.4b. They are parameterized simply by the product $c = \frac{1}{2}I\Omega$—e.g., in the signal dimension by $c_s = \frac{1}{2}I_s\Omega_s$. We observe that, as c increases, the desired spectral shape—given by f_1 in the band Ω—is increasingly tapered. The percent energy contained in the desired resolution cell I is given in Fig. 4.3: as c is increased, this becomes a seemingly very large percentage.

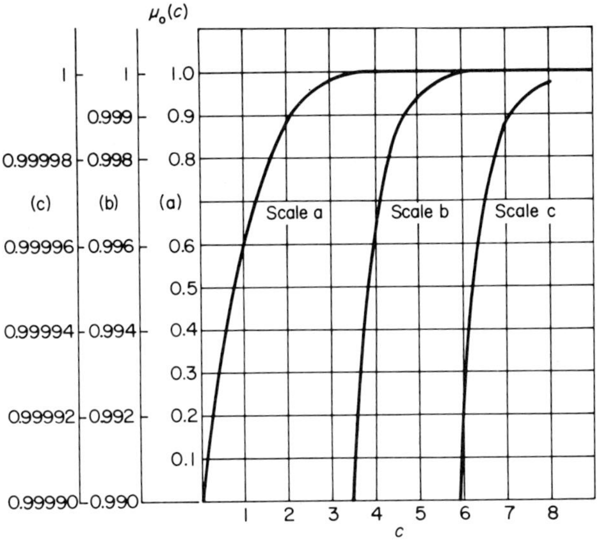

FIG. 4.3 Fraction of energy, $\mu_o(c)$, in desired resolution cell, I_s, as a function of time-bandwidth product, $c \equiv \frac{1}{2}I_s\Omega_s$, of signal modulation.

Hopefully, we set out to improve the sidelobe level of the system impulse response. In order to see what we have achieved, we must compute the impulse response (by computing the inverse Fourier transform of f_1 over the spectral band, Ω) and simply observe the relative sidelobe height. The family of impulse responses is shown in Fig. 8.5 and the relative height of the largest sidelobe in Fig. 8.13. Replotting Fig. 8.13 as Fig. 4.4, we see that as c increases

—i.e., as we increase the system bandwidth that we are willing to expend—the maximum sidelobe level does, in fact, decrease at a reasonably good rate and to desirably low levels.

The width ρ of the impulse response that is actually achieved is shown in Fig. 8.14, normalized by the interval I where we hoped to contain as much energy as possible; for the sake of illustration, the width ρ was defined to be the distance between half power points of the impulse response.

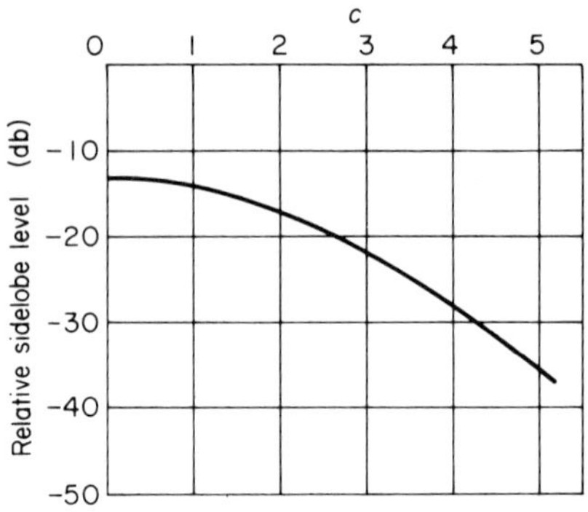

FIG. 4.4 Relative sidelobe level when system transfer function is selected from the prolate spheroidal wave function family (parameter c).

Thus, if $S_{00}(c; \omega)$ is a PSWF with parameter c, then we want the overall system transfer function

$$\tilde{F}(\omega)\tilde{B}_X(u) \cdot \tilde{H}(\omega, u) = S_{00}(c; \omega)S_{00}(c; u).$$

One might want to choose different c's in the two coordinate directions.

4.2.2. Least Squares Criterion

Although resolution criteria are very common in engineering work, especially when little a priori commitment is possible, a very common mathematical criterion when little a prior knowledge is available is least square error. That is, we try to choose a processing procedure that minimizes the integrated squared error; the processor is assumed linear, again because there is apparently no reason to assume otherwise

In our present application, we decide we wish to make a least squares estimate of the terrain—but we certainly are not literally interested in doing that. Rather we are generally interested in terrain structure of a certain size—not, e.g., in sizes of the order of a grain of sand, or even a leaf, or a city, or a continent. On the basis of the size of terrain structure that we are interested in we choose a *spectral window*, $W(\omega, u)$, say, that "preselects" the spectral part of the terrain that is of interest. Of course, making the selection of W more precise requires deciding what terrain structure is really of interest.

This criterion gives the optimum system directly: it is a LIV system with transfer function $W(\omega, u)$.

To this point, the discussion is rather trivial, but it is interesting to compare this point of view with the resolution criteria. First, suppose that we take $W(\omega, u)$ to be constant over some band: then we get the same processor we found with the "simple" resolution criterion (Sec. 4.2.1). But it generally is not reasonable that one would be equally interested in all terrain spatial frequencies to, say, ω, and not interested at all in the frequency $\omega + \varepsilon$, ε arbitrarily small! Second, in the light of the first observation, we decide that a tapered $W(\omega, u)$ makes more sense, reflecting a decreasing interest in higher spatial frequencies. But from the discussion of Sec. 4.2.1, this is roughly equivalent to deciding that the sidelobe level, a resolution criterion, should be smaller than the -13.4 db associated with a constant transfer function. In fact, if we decide to characterize an interest in relative frequencies of the terrain by selecting a member of the prolate spheroidal wave function family as a spectral taper, this is precisely equivalent to deciding what percent energy of the system impulse response is to be contained in the desired resolution cell!

Least squares estimation for specific objects. If we know specifically what terrain objects are of interest, we can require the processor to do best for those objects. Presumably, we do not know their number, orientations, or positions, else there would be little reason to form the image. Assuming for the moment that the desired terrain object G_1 is known, we wish to

$$\min_H \iint dt\, dx\, |G_1(t, x) - H(t, x) * Z(t, x)|^2$$

where H is the system impulse response and $G \equiv G_1 + G_c$ is the terrain. Note that we have assumed that the desired object G_1 is additive to the undesired part, G_c of G. It is easy to see that the best system has transfer function

$$\tilde{H} = \frac{1}{\tilde{F}\tilde{B}_X} \frac{\tilde{G}_1}{\tilde{G}_1 + \tilde{G}_c}$$

for (ω, u) where $\tilde{F}\tilde{B}_X \neq 0$. That is, we now have a rational procedure that specifies the spectral window $W(\omega, u)$ discussed immediately above.

We assume we know $\tilde{G}_1(\omega, u)$ within the phase due to the unknown orientation and location of the desired objects. But since the desired system is to give an image as an output, we do not wish to incorporate such phase information into the system filter anyway. The desired object is characterized by a reflectivity density that must either be computed—generally speaking an almost impossibly difficult task for typical objects of interest—or measured.

The background spectrum, $\tilde{G}_c$ must be determined in order to specify H. Of course, we do not know G_c unless indeed we measure it; that is, we would have to do something equivalent to forming an image with G_1 absent. For example, if it was known a priori that G_1 was not in a specific area, then we could compute G_c from such an area; this assumes a "homogeneity" of some kind over the area of interest; this is actually a reasonable procedure in some situations, but probably not in SAR systems.

A more reasonable procedure for SAR systems is to acknowledge that the clutter is not known and that the system that measured it would be too complex to incorporate into the SAR system. Rather, we attempt to characterize it as a random function (process) with a known spectral density. That is, we assume the spectral density has been measured a priori. Of course, modeling this clutter by a wide-sense stationary random process can only be of limited accuracy: the clutter process in mountains is presumably different than the clutter process in cities. We shall take up this point of view in Sec. 4.3.

4.3. CRITERIA FOR ADDITIVE NOISE

There has been a great deal of study of systems optimum under various criteria that acknowledge the unavoidable presence of additive noise. Models that include such noise are much better approximations to physical reality though certainly they often must be improved upon. In the present context, an orbital SAR system is likely to be modeled quite well by including additive noise, for such a system is likely to have limited emitted power capability.

We shall include the clutter in the additive noise when we are interested in specific objects on the terrain. Three common statistical criteria will be discussed: maximum signal-to-noise ratio ($\mathrm{SNR}_{\max}$), linear least square error (LMVE), and the method of maximum likelihood (ML).

4.3.1. Maximum SNR Criterion

The ordinary signal-to-noise (SNR) criteria are "local," or "point," criteria, in that they are concerned with the SNR at a particular point. In particular they place no value upon producing a faithful *image* of the object. This is not to say that approximate geometrical position is not preserved—for it can be. These criteria are in fact most sensibly regarded as *object detection* and *position estimation* criteria. Then it makes sense to devote the processor's design entirely to maximizing the signal due to the object at its geometrical position.

Suppose one has received a signal $S(x, y)$ along with additive, homogeneous noise $N(x, y)$ with spectral density $P_*(\omega, u)$: then it is well known that the linear, invariant, filter H which maximizes the SNR [at $(x, y) = (0, 0)$] has transfer function

$$\tilde{H}(\omega, u) = \frac{\bar{\tilde{S}}(\omega, u)}{P_*(\omega, u)} \qquad (4.1)$$

and

$$\mathrm{SNR}_{\max} = \frac{1}{4\pi^2} \iint \frac{|\tilde{S}(\omega, u)|^2}{P_*(\omega, u)}\, d\omega\, du. \qquad (4.2)$$

We can apply this result to the SAR system design in several cases.

First, suppose that we have a point-scattering object so that the received signal is $F(t)B_X(x)$, and that the noise is white (amplitude N_o) over the spectral band of FB_X—as e.g., would be the case if the noise is entirely N_1. Then Eq. (4.1) now gives

$$\tilde{H}(\omega, u) = N_o^{-1}\bar{\tilde{F}}(\omega)\bar{\tilde{B}}_X(\omega), \qquad (4.1a)$$

the *matched filter*, and Eq. (4.2) now gives

$$\mathrm{SNR}_{\max} = \frac{E_s}{N_o}, \qquad (4.2a)$$

a ratio of signal energy to white noise spectral density. These are famous results.

Second, suppose that we have a specific object with reflectivity density G_1 that we wish to detect and determine its position; further, we assume that the background can be modeled by a homogeneous random field with spectral density P_c. Then we now have

$$\tilde{S}(\omega, u) = \tilde{G}_1(\omega, u)\tilde{F}(\omega)\tilde{B}_X(u)$$

and

$$P_*(\omega, u) = P_c(\omega, u)\,|\tilde{F}(\omega)\tilde{B}_X(u)|^2 + P_n(\omega, u),$$

where P_n is the spectral density of all additive noises—the usual RF noises, recorder (e.g., film grain) noises, etc. Now Eqs. (4.1) and (4.2) give, respectively,

$$\tilde{H}(\omega, u) = \frac{\bar{\tilde{G}}_1(\omega, u)\bar{\tilde{F}}(\omega)\bar{\tilde{B}}_X(u)}{P_c|\tilde{F}\tilde{B}_X|^2 + P_n}$$

and

$$\mathrm{SNR}_{\max} = \frac{1}{4\pi^2} \iint \frac{|\tilde{G}_1\tilde{F}\tilde{B}_X|^2}{P_c|\tilde{F}\tilde{B}_X|^2 + P_n}\, d\omega\, dv. \qquad (4.2b)$$

We note the following important facts.

Property 4.1. "The maximum signal-to-noise ratio $\mathrm{SNR}_{\max}$ is independent of the phase of $\tilde{F}\tilde{B}_X$."

This implies, e.g., that the method employed in the SAR system for achieving resolution in azimuth does not carry any penalty in signal-to-(additive) noise ratio. We have the same implication with respect to the use of dispersed modulations, F, in the range channel.

Property 4.2. "If the spectral density of the additive noise is constant over the spectral band of the signal and does not depend on the signal, then the $\mathrm{SNR}_{\max}$ depends on the signal only through its energy."

Corollary 4.2.1. "If Prop. 4.2 is true, then the signal $F(t)B_X(x)$ can be chosen to meet simultaneously a signal-to-noise restraint and a resolution requirement."

For example, we could choose $\tilde{F}$ and $\tilde{B}_X$ to be a PSWF, as discussed in Sec. 4.2, to achieve a resolution requirement defined in terms of energy.

We note that a specified object against a clutter background, as we have modeled it, results in a signal-dependent additive noise spectral density, and the $\mathrm{SNR}_{\max}$ explicitly depends on $|\tilde{F}\tilde{B}_X|$. We have however, the following limiting case.

Property 4.3. "If in Eq. (4.2b), $P_c|\tilde{F}\tilde{B}_X|^2 \gg P_n$ for (ω, u) in the spectrum of $|\tilde{F}\tilde{B}_X|^2$, then the $\mathrm{SNR}_{\max}$ is independent of the signal modulation."

This limiting case, is in fact, to be expected unless the object of interest is an extremely strong scatterer, for its signal at any time is simultaneously received with all the myriad signals of objects being illuminated at that instant.

We noted above that when the additive noise spectral density depends upon the signal modulation, so does $\mathrm{SNR}_{\max}$. Thus one may wish to choose the signal modulation FB_X so as to further maximize the signal-to-noise ratio.

Property 4.4. "Let $\Sigma \equiv \{(\omega, u)\colon -\Delta < \omega < \Delta,\ -\Omega < u < \Omega\}$. Assume that $|\tilde{G}_1|^2$, P_n, and P_c are all positive for $(\omega, u) \in \Sigma$, and set $f \equiv |\tilde{F}\tilde{B}_X|^2$. Then $\mathrm{SNR}_{\max}$ (Eq. 4.26) is maximized over f, subject to the restraint

$$\frac{1}{4\pi^2} \iint\limits_{\Sigma} f(\omega, u)\, d\omega\, du = \mathscr{E},$$

by

$$f = \max\left\{0, \frac{1}{P_c}\left[\left(\frac{|\tilde{G}_1|^2 P_n}{\lambda}\right)^{1/2} - P_n\right]\right\}, \qquad (\omega, u) \in \Sigma,$$

where λ is determined by the energy restraint."

Proof: We equivalently choose f to maximize the (Lagrangian) form:

$$\mathrm{SNR}_{\max} - \lambda\mathscr{E} = \frac{1}{4\pi^2} \iint_{\Sigma} \left(\frac{|\tilde{G}_1|^2 f}{P_c f + P_n} - \lambda f \right) d\omega\, du.$$

Clearly we choose f ($f \geq 0$) so as to maximize the integrand at each point $(\omega, u) \in \Sigma$. It is easy to see that there is a maximum at

$$f_* = \frac{1}{P_c} \left[\left(\frac{|\tilde{G}_1|^2 P_n}{\lambda} \right)^{1/2} - P_n \right]$$

and that if $f_* < 0$ then $(\mathrm{SNR}_{\max} - \lambda\mathscr{E})$ is monotone decreasing for $f \geq 0$, so that then $f = 0$ gives the maximum, $f \geq 0$. Q.E.D.

The special case of a point scattering object ($|\tilde{G}_1|^2 = a$, constant) and white noise ($P_n = N_o$, constant) is worth noting. Since now

$$\alpha \equiv \left[\left(\frac{aN_o}{\lambda} \right)^{1/2} - N_o \right]$$

must be a positive constant ($\iint f = \mathscr{E}$) now

$$f = \frac{\alpha}{P_c}, \qquad (\omega, u) \in \Sigma.$$

Further, the system transfer function is

$$\begin{aligned} \tilde{H} f &= \frac{\alpha}{\alpha + N_o} \cdot \frac{1}{P_c} \\ &= \left[1 - \left(\frac{\lambda N_o}{a} \right)^{1/2} \right] \frac{1}{P_c}. \end{aligned} \tag{4.3}$$

The form of the result is reasonable: the spectrum of the signal modulation is emphasized where the clutter spectral density is small. This result can be unattractive from a resolution criterion point of view. For example, suppose P_c is unimodal about $(\omega, u) = (0, 0)$: then $|\tilde{F}\tilde{B}_X|^2 \tilde{H}$ has "emphasized edges," and this results in high sidelobes in the system impulse response. Further, such a system transfer function is quite the opposite from what is desirable when we use the "percent energy" resolution criterion.

4.3.2. Least Squares Criterion

A model of general terrain that is useful in imaging system design assumes that the reflectivity density is a homogeneous, random field (with spectral density P_g). We are generally interested, as we have pointed out above, in only a selected spatial frequency band of this terrain process: we have discussed how explicitly or implicitly such a "spectral window, W," might be chosen. The part of the terrain process of interest then has spectral density $P_{gW} \equiv P_g|W|^2$. Referring to Fig. 4.2, the random field available for processing has spectral density

$$|\tilde{F}(\omega)\tilde{B}_X(u)|^2 P_{gW}(\omega, u) + P_n(\omega, u); \tag{4.4}$$

we assume also that it has zero mean.

A reasonable aim for the system would be recovering the reflectivity density, as well as possible, for subsequent processing such as viewing by a human. By "as well as possible" we shall mean "with least square error by a linear filter." (The filter will necessarily be quasi-invariant, as we have pointed out.) One of the principal advantages of this criterion is that it requires only the a priori knowledge of spectral density—as little as we could hope for.

It is well known that the filter H represented in Fig. 4.2 should have transfer function

$$\tilde{H}(\omega, u) = \frac{\bar{\tilde{F}}(\omega)\bar{\tilde{B}}_X(u) P_{gW}(\omega, u)}{|\tilde{F}(\omega)\tilde{B}_X(u)|^2 P_{gW}(\omega, u) + P_n(\omega, u)} \tag{4.5}$$

and the resulting minimum mean square error

$$\text{MSE} = \frac{1}{4\pi^2} \iint \frac{P_{gW} P_n}{|\tilde{F}\tilde{B}_X|^2 P_{gW} + P_n} \, d\omega \, du. \tag{4.6}$$

An obvious comparison of Eqs. (4.5) and (4.1b) can be made; the similarities in the filters, optimum under these two criteria, accounting for additive noise is striking. We also note that the minimum MSE does not depend on the phase of $\tilde{F}\tilde{B}_X$—just as the maximum SNR [Eq. (4.2b)] did not—and therefore the SAR's method of gathering data in a "dispersed form" in the azimuth channel does not impose any cost in MSE.

We see from Eq. (4.5) that for very strong additive noise,

$$P_n \gg |\tilde{F}\tilde{B}_X|^2 P_{gW},$$

$$\tilde{H} \to \bar{\tilde{F}}\bar{\tilde{B}}_X \frac{P_{gW}}{P_n}.$$

If we assume P_{gW} to be constant over the band of $|\tilde{F}\tilde{B}_X|$, then we have the same filter we obtained under the maximum SNR criterion for a point scatterer: the analogy between a "white" P_{gW} and a point scatterer is intuitively reasonable.

Signal spectrum modulus optimization. We can choose the signal spectrum modulus so as to further reduce the MSE of Eq. (4.6).

Property 4.5. "Let

$$\Sigma \equiv \left\{(\omega, u): \quad \begin{matrix} -\Delta < \omega < \Delta \\ -\Omega < u < \Omega \end{matrix}\right\}.$$

Assume that P_n and P_{gW} are positive for $(\omega, u) \in \Sigma$ and set $f \equiv |\tilde{F}\tilde{B}_X|^2$. Then MSE [Eq. (4.6)] is minimized over f, subject to the restraint

$$\frac{1}{4\pi^2} \iint_{\Sigma} f(\omega, u)\, d\omega\, du = \mathscr{E},$$

by

$$f = \max\left\{0, \left[\left(\frac{P_n}{\lambda}\right)^{1/2} - \frac{P_n}{P_{gW}}\right]\right\}, \qquad (\omega, u) \in \Sigma, \tag{4.7}$$

and

$$\min_f \text{MSE} = \frac{1}{4\pi^2} \iint_{\Sigma_1} (\lambda P_n)^{1/2}\, d\omega\, du + \frac{1}{4\pi^2} \iint_{\Sigma_2} P_{gW}\, d\omega\, du, \tag{4.8}$$

where $\Sigma_2(\Sigma_2 \subset \Sigma)$ is the (ω, u)-set where $f = 0$, and Σ_1 is the rest of Σ. The energy restraint $\mathscr{E}$ determines λ."

Proof: We equivalently choose f to minimize the (Lagrangian) form

$$\text{MSE} + \lambda\mathscr{E} = \frac{1}{4\pi^2} \iint_{\Sigma} \left(\frac{P_{gW} P_n}{P_{gW} f + P_n} + \lambda f\right) d\omega\, du.$$

Clearly we choose f $(f \geq 0)$ so as to minimize the integrand at each point $(\omega, u) \in \Sigma$. It is easy to see that there is a minimum at

$$f_* = \left(\frac{P_n}{\lambda}\right)^{1/2} - \frac{P_n}{P_{gW}},$$

and that if $f_* < 0$ then $(\text{MSE} + \lambda\mathscr{E})$ is monotone increasing for $f \geq 0$, so that then $f = 0$ gives the minimum, $f \geq 0$. Q.E.D.

We note that the overall system transfer function becomes

$$\tilde{H}\tilde{F}\tilde{B}_X = \begin{cases} 0, & (\omega, u) \in \Sigma_2, \\ \left[1 - \dfrac{(P_n \lambda)^{1/2}}{P_{gW}}\right]^2, & (\omega, u) \in \Sigma_1. \end{cases} \tag{4.9}$$

Example. In a properly designed system, the signal-to-noise ratio is large in the sense that

$$|\tilde{F}\tilde{B}_X|^2 P_{gW} \gg P_n, \qquad (\omega, u) \in \Sigma_1. \tag{4.10}$$

It is easy to see that Eq. (4.10) implies both

$$P_{gW} \gg (P_n \lambda)^{1/2}$$

and

$$\left(\frac{P_n}{\lambda}\right)^{1/2} \gg \frac{P_n}{P_{gW}}.$$

Therefore, both the signal modulation spectrum modulus and the overall SAR transfer function become approximately constant over the spectral band. Thus, in this limiting case we are led back to the same system we obtained with the "simple minded" resolution interior first discussed.

Additionally, it is usually true that the additive noise has an approximately constant spectral density over the spectral band of interest: $P_n \to N_o$. Further, suppose that

$$P_{gW}(\omega, u) \approx P_{go}[1 - \varphi(\omega, u)],$$

where φ might represent a tapered weighting of some kind, $\varphi_{\max} \ll 1$. Then Eq. (4.9) gives

$$\tilde{H}\tilde{F}\tilde{B}_x \approx 1 - \frac{\sqrt{N_o \lambda}}{P_{go}} \varphi(\omega, u).$$

The overall system has the same functional weighting as P_{gW} but much reduced in amplitude.

4.3.3. Maximum Likelihood Criterion

One of the most common methods of detecting signals and estimating unknown signal parameters is the method of maximum likelihood. The basic idea is that it is reasonable to form the ratio f of the probability of occurrence of the observed data, say z, assuming signal (object) present, to the same probability assuming the signal (object) absent,

$$f(z) \equiv \frac{\text{Prob}\{z \mid \text{object present}\}}{\text{Prob}\{z \mid \text{object absent}\}},$$

and to compare this ratio (a random variable) to unity: if $f(z) \geq 1$, we decide "object present," and if $f(z) < 1$, we decide "object absent." The ratio f is called the "likelihood ratio." It is also reasonable to estimate the values of the unknown parameters of the signal (object) by maximizing $f(z)$ with respect to these parameters; these parameter estimates are called "maximum likelihood estimates (MLE)."

Leaving aside for the moment the meaning of the merely suggestive expression for $f(z)$ above, we note that, beyond the reasonableness of the maximum likelihood procedure, it can be shown that this procedure has other desirable properties.

For example, if the a priori probabilities of the two events, {object present} and {object absent}, are equal, then the maximum likelihood test described above also minimizes the probability of error. It can be shown that for a variety of other detection criteria (Bayes, minimax, Neyman-Pearson) that the desired test statistic is always $f(z)$, the likelihood ratio, and that only the threshold with which $f(z)$ is compared need be varied to accommodate these various criteria. This fact lends great weight to having the processor structure realize the computation $f(z)$.

The maximum likelihood method of estimating parameters also has desirable properties. E.g., it can be shown that any unbiased estimator of an unknown parameter has a lower bound on its (error) variance, the so-called Cramer-Rao lower bound: an unbiased estimator that achieves this bound is termed "efficient." Efficient estimators do not necessarily exist in a particular problem, but if an efficient estimator exists, then it is a maximum likelihood estimator. Further, it can be shown that under reasonable conditions the maximum likelihood estimators are "asymptotically efficient"—e.g., as the signal-to-noise ratio increases in the specific problem we will discuss here—and are "consistent"—that is, converge in probability to the true parameter values.

Returning to the question of interpreting the formal expression for $f(z)$, suppose first of all the observation is—as is not true for us here—a single random variable with a probability density function under each hypothesis; then $f(z)$ is just the ratio of these densities. Trivially we generalize to the case when $z \equiv (z_1, \ldots, z_n)$. But here the data is a sample function of a random field—by definition an uncountably infinite family of random variables; we cannot write down a joint density function for such a family, and, in fact, the probability of observing any particular sample function can be zero. The way around the uncountably infinite difficulty is to sample in a suitable way and take a suitable limit as the number of samples becomes large; the difficulty of zeros is avoided by forming the ratio of densities prior to the limit operation. These superficial remarks do not do justice to the difficulties involved; in fact, with few exceptions, rigorous results are restricted to Gaussian processes, and even formal practical use of the procedure is severely restricted by the difficulty of finding the likelihood functional $f(z)$.

The problem to which we restrict our attention in this section is the detection, with estimation of unknown parameters, of an otherwise completely known signal (due to an object of interest) received along with additive, Gaussian noise.

If the spectral density of the additive Gaussian noise is constant (of value N_o) over the spectrum of the signal, then it is well known that the likelihood ratio, or functional, is

$$f_1(z) = \exp\left\{\frac{1}{N_o} \iint_A z(t, x) g_1(t, x; \alpha)\, dt\, dx - \frac{1}{2N_o} \iint_A [g_1(t, x; \alpha)]^2\, dt\, dx\right\}$$

where $z(t, x)$, $(t, x) \in A$ is the data received in the aperture $A \equiv T \times X$ and g_1 is the signal known except for parameter α. The optimum receiver is defined by this equation; it calculates $f(z)$, maximizes it with respect to α, and compares it to a threshold—which can be adjusted according to which criteria is used, and which is unity if a maximum likelihood test is made. The value(s) of α that maximize $f_1(z)$ are, by definition, the maximum likelihood estimates of the unknown parameters.

We note that the first term of the exponent is a linear function of the data; the processor is linear and, in fact, performs the matched-filter operation discussed in Sec. 4.3.1 that resulted there in the signal-to-noise ratio criterion and a processor assumed linear, a priori.

Here we get more, however, for we have a procedure to estimate the object's unknown parameters. Generally in the present situation, a minimal set of unknown parameters is the object's amplitude, phase, and position. At this point, it is convenient to change to complex representations for the signal and data; for then, the maximization of $f_1(z)$ over the unknown amplitude and phase is very easy. Therefore, if G and Z are the complex representations of g and z, respectively, we can show that

$$f_1(z) = \exp\left\{-\frac{1}{2N_o} \operatorname{Re} \iint \bar{G}_1(t, x; \alpha) Z(t, x) + \frac{1}{4N_o} \iint |G_1(t, x; \alpha)|^2\right\}.$$

Now we set

$$G_1(t, x; \alpha) \equiv \tilde{b}\, G_{11}(t - \tilde{t}, x - \tilde{x})$$

where $\tilde{b}$ is the unknown complex amplitude of the signal and $(\tilde{t}, \tilde{x})$ its unknown position. Note that we can and will maximize $\log f_1(z)$.

It is now easy to show that

$$\max_{\tilde{b}} \log f_1(z) = \frac{|\iint \bar{G}_{11}(t - \tilde{t}, x - \tilde{x}) Z(t, x)\, dt\, dx|^2}{\iint |G_{11}(t, x)|^2\, dt\, dx} \tag{4.11}$$

and the MLE

$$\hat{b} = \frac{\iint G_{11}(t - \hat{t}, x - \hat{x}) Z(t, x)\, dt\, dx}{2\iint |G_{11}(t, x)|^2\, dt\, dx},$$

where the MLE's of $\tilde{t}$ and $\tilde{x}$ have been denoted $\hat{t}$ and $\hat{x}$. We have assumed that for small variations of the "test" position $(\tilde{t}, \tilde{x})$ the integral

$$\iint |G_{11}(t - \tilde{t}, x - \tilde{x})|^2$$

is independent of $(\tilde{t}, \tilde{x})$. We note that the magnitude of the MLE of signal amplitude, $|\hat{b}|$, is proportional to $\max/\tilde{b} \log f_1(z)$. Usually, we are only interested in relative magnitudes anyway; if this is the case, we have the following simple processor:

> We pass the data $Z(t, x)$ through a linear filter represented by the complex impulse response $\bar{G}_{11}(-t, -x)$, and we search over the two-dimensional output of this filter to find its maximum: the value of this maximum is proportional to the MLE $|\hat{b}|$, and the location of this maximum is the MLE's $\hat{t}$ and $\hat{x}$.

Generally, the unknown signal phase, Arg b_o, is of no interest.

This is a conceptually very simple processor and is the one whose realization by a coherent optical system will be discussed at length in Chapter V.

As we have noted many times above, the possible objects of actual interest lie in a background, or clutter, and therefore the application of the results just derived are not obvious. One possibility is to assume that the objects of interest are point scatterers. One is "saved" to a great extent, because the processor turns out to be linear; thus, a superposition of point scatterers and, by tenuous extension, distributed scatterers is processed in an apparently reasonable way.

There are other points of view to take. One is to model the background by a Gaussian process and rederive the likelihood ratio. The defect is obvious: we are rather in the dark about when, if ever, such an assumption about the terrain is reasonable. Proceeding nonetheless, one can show that now the likelihood ratio is

$$f_2(z) = \exp\left\{\iint_A z(t, x)q(t, x; \alpha)\, dt\, dx - \frac{1}{2}\iint_A q(t, x; \alpha)g_1(t, x; \alpha)\, dt\, dx\right\},$$

where q is given as the solution of the integral equation

$$g_1(t, x; \alpha) = \iint_A R_n(t - t_1, x - x_1)q(t_1, x_1; \alpha)\, dt_1\, dx_1, \qquad (t, x) \in A.$$

Generally, such integral equations cannot be solved exactly. There are techniques when the spectral density, P_n of the noise is rational, but they can be messy. To get an idea of the effect of colored noise on ML processor, suppose that the correlation function R_n of the noise is zero, outside an area small compared to A, the processing aperture area. Then rather small error is incurred over most of the aperture, if we simply set A equal to the whole plane: now we can solve the integral equation using the Fourier transform, and we find that the system transfer function

$$\tilde{q}(\omega, u) \approx \frac{\tilde{g}_1(\omega, u)}{P_n(\omega, u)}.$$

This is the same system impulse response that was found under the $SNR_{\max}$ criterion, under similar assumptions on the size of the processing aperture A!

We repeat that we can employ this result for clutter, by taking $P_n = P_1 + P_c$, where P_1 is the spectral density of the system additive noise, and P_c is the spectral density of the background, or clutter.

4.4. COMPARISON OF CRITERIA

It is of interest now to compare the system impulse responses that have been found optimum under various criteria and conditions. The most significant thing is that the optimum systems, under criteria that do not account for noise of any kind, are in at least rough agreement with the systems that are optimum with respect to criteria that take into account additive noise, random or not. In part, this is due to the linearity assumption explicit in all but one criterion. But it is easy to see how most of these criteria are sensible relative to each other, on the basis of their own criteria.

One major result is, that the restriction to *linear* systems with all the criteria except that of maximum likelihood appears to be a less serious restriction, since the maximum likelihood criterion resulted in a linear optimum system (under the assumption of known signal in additive Gaussian noise). We cannot be completely satisfied with this result, however, because we are doubtful about modeling the terrain by a Gaussian process.

The fact that the $SNR_{\max}$ criterion—which assumed a linear system but no Gaussian process—results in the same system as the maximum likelihood criterion—which assumed a Gaussian process but not a linear system—gives weight to selecting this system when faced with additive noise as the major error source. The characterization of the system does require a specification of the spectrum of the object of interest. This, of course, can be difficult, but is the price to be paid for improved system performance.

We have discussed the design of the spectrum modulus of the complex signal modulation, in addition to the processor design, under the maximum SNR and MSE criteria. Both typically result in a tapered weighting to said modulus. In Chapter VIII, we shall assume a processor such as found here—

specifically the matched filter—but design this modulus from the point of view of suppressing ambiguities. There we shall find that tapered weightings are desirable, giving a rough agreement with the results of this chapter.

APPENDIX 4A: SPECTRAL DENSITY OF TWO-DIMENSIONAL SYSTEM NOISE

Suppose we let n_t be a sample function of the noise that is always present in a microwave radar set: it is a sample function of a zero-mean Gaussian process, white over the spectral band of interest. In the SAR system, segments of this noise (of time duration equal to the range delay interval being recorded) are recorded sequentially in time "side-by-side," to form the usual two-dimensional presentation. It is a very good approximation to assume that these resulting "noise strips" are mutually independent. We need only be concerned with this two-dimensional (Gaussian) noise process after it has been smoothed by a filter q, say. Thus, analytically, the two-dimensional noise is

$$\begin{aligned} n_1(t, x) &\equiv \{\sum n_k(t)\, \delta(x - kX_s)\} * q(x) \\ &= \sum n_k(t) q(x - hX_s) \end{aligned}$$

where the smoothing filter q is such that

$$\tilde{q}(u) = \begin{cases} 1, & |u| < \tfrac{1}{2}u_s, \\ 0, & |u| > \tfrac{1}{2}u_s, \quad u_s \equiv \dfrac{2\pi}{X_s}. \end{cases}$$

In order to apply the results on system theory discussed above, we need to know the spectral density or, equivalently, the covariance function of the random process $\{n(t, x)\}$. We compute

$$\begin{aligned} R_{n_1}(t_1, x_1; t_2, x_2) &\equiv E\{n_1(t_1, x_1) n_1(t_2, x_2)\} \\ &= R_n(t_1 - t_2) \sum q(x_1 - kX_s) q(x_2 - kX_s). \end{aligned}$$

Examining the sum, $\mathscr{S}$, we write both q's in their Fourier integral representation:

$$\mathscr{S} = \frac{1}{4\pi^2} \int du\, e^{iux_1} \tilde{q}(u) \int dv\, e^{ivx_2} \tilde{q}(v) \sum e^{-i(u+v)kX_s}.$$

We replace the latter sum using the identity

$$\sum e^{-i(u+v)kX_s} = u_s \sum \delta(u + v - nu_s),$$

and find therefore

$$\mathcal{S} = \frac{u_s}{4\pi^2} \sum \int du \exp[i(x_1 - x_2)u - inu_s X_s] \tilde{q}(u) \tilde{q}(-u + nu_s)$$

$$= \frac{u_s}{4\pi^2} \int du \exp[i(x_1 - x_2)u] \, |\tilde{q}(u)|^2.$$

Thus finally

$$R_{n_1}(t_1, x_1; t_2, x_2) = R_n(t_1 - t_2) \cdot \frac{u_s}{4\pi^2} \int du \exp[i(x_1 - x_2)u] \, |\tilde{q}(u)|^2.$$

$\{n_1(t, x)\}$ is therefore a homogeneous two-dimensional field.

The spectral density S_s of $\{n_1\}$ is therefore

$$S_{n_1}(\omega, u) = S_n(\omega) \cdot \frac{u_s}{2\pi} \, |\tilde{q}(u)|^2.$$

Since S_n is of constant value N_o over all ω of interest, and

$$\tilde{q}(u) = \begin{cases} 1, & |u| < \tfrac{1}{2}u_s, \\ 0, & |u| > \tfrac{1}{2}u_s, \end{cases}$$

we have

$$S_{n_1}(\omega, u) = \begin{cases} \dfrac{u_s}{2\pi} N_o, & |u| < \tfrac{1}{2}u_s \\ 0, & |u| > \tfrac{1}{2}u_s, \quad \text{all } \omega. \end{cases}$$

As is quite reasonable, the spectral density of the two-dimensional process is proportional to both N_o and the spatial rate u_s at which the "one-dimensional" noise samples are recorded.

APPENDIX 4B: IMAGE CALCULATION BY METHOD OF GEOMETRICAL OPTICS

We have at various points in Chapter IV discussed processing procedures for "known" objects. We wish to point out here some fairly obvious—yet frequently forgotten—facts about images produced by SAR systems: the "known" object can produce an image quite different from an ordinary photographic or visual image. We have already discussed the method of physical optics which can often describe the image: here we wish to use the simpler method of geometrical optics.

When the major dimensions of an object and its radii of curvature are large relative to a wavelength, the method of geometrical optics is the most efficient way to calculate an approximation to the scattered field. The method is based simply on Snell's law.

The application of this method is especially easy when we assume, without loss of essentials, that the data processor is an optical system. *For then, "rays" can be graphically traced from object plane to image plane for the entire SAR system.* In fact, ray tracing methods are almost exclusively used in the design of optical systems.

As an example that shows some typical behavior for this class of objects, consider the aircraft shown in Fig. 4.5, viewed by the radar set over a narrow range of angles centered on the angle from which the photograph was taken. Without going into the complexities of the geometry, we may readily see that

FIG. 4.5 Photographic image, in ordinary light, of an object approximately smooth relative to the wavelength of the illumination.

FIG. 4.6 Radar image of the object of Fig. 4.5, simulated by incoherent, unidirectional illumination. (The leading edge of the tail, a planar surface to optical waves, would be much less reflective at radar wavelengths.)

there are only a small number of regions on the object for which an emitted ray will be reflected directly back to the radar set (Fig. 4.6): *these are the regions, called "specular" regions, whose surface normals, if extended, would pass through the synthetic aperture.* If the wavelength is small relative to the size of the aircraft and to the viewed radii of curvature, and if the resolution is finer than the separation of at least some of these regions, the following conclusions are fairly obvious:

1. The image of the object will be composed of a finite number of "spots" —termed "hot-spots" in the radar trade—and which we will call "specular images."
2. If there are N regions of the type described, and M of them are separated by the resolution distance, the image will be composed of M specular images.
3. Making the resolution finer does not basically change the image: it can at most cause some of the specular images to divide.
4. The image is independent of wavelength over the range of λ that are small relative to the dimensions and radii of curvature of the object, and for which the object is still smooth.
5. The image of broadband waveforms does not change from the monochromatic image within the range of λ described above and hence *diversity of λ within this range will not change the image.*
6. If the object is still smooth at the wavelength of light, an ordinary "flash photograph" of the object would give the same image. ("Ordinary daylight photographs" of an object such as an aircraft are not composed of numerous "spots" in part because they are slightly rough, but probably mostly because they are actually illuminated from many directions by the ambient light due to the multiple scattering of sunlight.)
7. It can be estimated from a rather casual examination of the object whether or not angle diversity will change the image; the most probable effect is simply that the number of spots would change with the aspect angle.
8. Since the specular images do not depend upon polarization, polarization diversity will not change the image.

It is also quite clear that there is little possibility that the object itself could be uniquely determined in such a situation. For we could obviously distort the surface of the body that was not contributing rays to the image in an arbitrary way, as long as we kept the radii of curvature large compared to a wavelength and did not create new contributing regions—we could even subdivide the object into as many parts as there were spots. All this leaves the image predicted on the basis of geometrical optics unchanged.

It is therefore apparent that *the identification of such scatterers involves a recognition of the geometrical arrangement of the specular images* based upon a priori knowledge—an operation most feasibly performed by a human being at the present time.

It is to be emphasized that the basic nature of the image of this type of object has almost nothing unique about it due to the SAR technique, coherent illumination, and system resolution (except of course as to the resolution of the specular images).

COMMENTS AND REFERENCES

The discussion of the maximum SNR, minimum MSE, and ML method criteria is an application of well-known and standard communication theory results. See, e.g., Davenport and Root (1958), and Brown (1963b). Resolution criteria have been summarized at some length by Brown (1963b); e.g., resolution has been discussed via the method of maximum likelihood by Root (1962). The optimization of the signal spectrum modulus under MSE criterion was given by Brown and Palermo (1963a).

An interesting variant on the MSE approach was obtained by W. M. Brown, R. B. Crane, and J. G. Wendel; see Brown *et al.* (1964). The problem is to minimize the MSE in estimating the *modulus* of the reflectivity density: this is apparently more desirable than estimating the reflectivity density itself because the final step in data processing—perhaps the eye-brain combination—is an operation that responds only to the modulus of the image. The Gaussian assumption is made for the reflectivity density and additive noise—but not linearity of the processor—and it is found that the optimum procedure is, essentially, to use the processor found above assuming linearity followed by an operation that takes the square-modulus.

For a discussion of the usual "RF noise" see, e.g., Barton, in Berkowitz (1965).

V

Optical Processing

5.1. ELEMENTS OF OPTICAL PROCESSING

It is satisfying that, with essentially only the Kirchhoff-Huygens diffraction integral as a point of departure, we are able to completely explain rather simply the synthesis of the optical systems that we shall need here. We shall be able to derive the imaging, Fourier transforming, and integrating operations that are, in principle, sufficient to realize any LIV system. We, of course, content ourselves with describing the ideal operations of optical systems; we will not enter into the arcana of the centuries-old optical art that exists in part to control the nonideal effects in optical systems.

5.1.1. Kirchhoff-Huygens Principle

The Kirchhoff-Huygens diffraction integral (Born and Wolf (1965), Chapter VIII, esp. Sec.8.3) is the only result from physics that we need to describe the operations of the coherent optical processors we shall discuss here. This is because all dimensions of interest of the optical system are very large relative to a wavelength, and hence boundary effects are small. Further, because there are no polarization effects, we need only be concerned with the scalar field E within the optical system.

Referring to Fig. 5.1 we denote the scalar field of the wave, propagating to the right, by $E(x, y, z)$. Huygens' principle asserts that the scalar field at any point to the right ($x > x'$) is the sum of spherical waves propagating from every point (x', y', z') with initial intensity $E(x', y', z') \equiv E_{x'}(y', z')$. Kirchhoff derived a corresponding expression from the scalar wave equation directly, with certain assumptions about boundary conditions—essentially that, referring to Fig. 5.1, the field $E_{x'}(y', z')$ in the aperture is the same as it would be if no screen were present at $x = x'$, and that the field $E(0+, y, z)$ on the shadow ($x = 0+$) side of the screen is zero. The result is, in complex representation,†

$$E(x, y, z) = \frac{-ik}{4\pi} \iint_A \frac{\exp(ik|\mathbf{r} - \mathbf{r}'|)}{|\mathbf{r} - \mathbf{r}'|} E_{x'}(y', z') \cdot \tfrac{1}{2}[1 + \cos\chi(\mathbf{r} - \mathbf{r}')]\, dy'\, dz' \tag{5.1}$$

where the "obliquity" factor $\frac{1}{2}[1 + \cos\chi(\mathbf{r} - \mathbf{r}')]$ accounts for the direction of propagation.‡ We shall denote $E(x, y, z)$ by $E_x(y, z)$, $x > 0$, since we are generally interested in the field at various planes, $x =$ constant in the optical system.

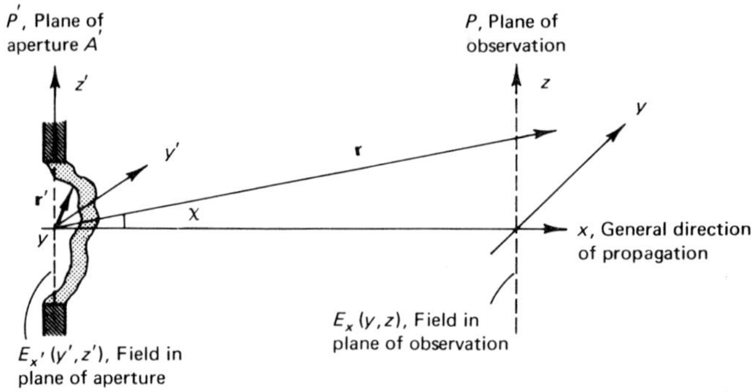

FIG. 5.1 Geometry for calculation of Kirchhoff–Huygens diffraction integral.

We may simplify this integral for our purposes, because of the restricted angle through which the field is propagated from the plane $x = 0$. As we shall see (Sec. 5.1.3), the field is generally created by illuminating a "transparency" with a uniform plane wave propagating in the x-direction. The "transparency" is, e.g., an exposed piece of ordinary photographic film, i.e., a negative: such film—or any other recording medium—has a finite spatial bandwidth, or finest resolvability. For practical recording media, this is less than 10^3 cycles (lines)/mm $= 10^6$ cycles/m; therefore, the "smallest" spatial variation possible is of the order of 10^{-6} m. Thus, as can be seen from Eq. (5.1), although it is well known, such a structure will scatter the wave through an angle $\lambda/10^{-6} \approx 10^{-1}$ rad, much as an antenna of that size (10^{-6} m) would propagate the wave mainly into an angle $\lambda/10^{-6}$ rad.

Hence, in our applications we may take the obliquity factor equal to one as a good approximation; and, if the x-axis is approximately aligned with

† That is, the scalar field is given by $\text{Re}\{E(x, y, z)e^{i\omega_0 t}\}$.
‡ χ is the angle between $(\mathbf{r} - \mathbf{r}')$ and the normal to the wavefront.

this scattering cone, we may use the approximation

$$|\mathbf{r}-\mathbf{r}'| = [(x-x')^2 + (y-y')^2 + (z-z')^2]^{1/2}$$
$$\approx (x-x') + \frac{(y-y')^2 + (z-z')^2}{2(x-x')}, \tag{5.2}$$

the so-called "*sagittal approximation*" for the phase of $(\exp ik|\mathbf{r}-\mathbf{r}'|)/|\mathbf{r}-\mathbf{r}'|$; for the latter's amplitude approximation we use

$$|\mathbf{r}-\mathbf{r}'| \approx |x-x'|$$
$$= (x-x'), \qquad x > x'. \tag{5.3}$$

Substituting Eqs. (5.2) and (5.3) into Eq. (5.1) we have then

$$E_x(y, z) \approx \frac{-ik}{4\pi}\frac{\exp[ik(x-x')]}{(x-x')} \iint_{A'} \exp\left[ik\frac{(y-y')^2 + (z-z')^2}{2(x-x')}\right]$$
$$\cdot E_{x'}(y', z')\, dy'\, dz'. \tag{5.4}$$

Equation (5.4) is the basic relation between the fields at two planes P' and P, with coordinates (x', y', z') and (x, y, z), respectively, for $x > x'$, and using the sagittal approximation in the Kirchhoff-Huygens diffraction integral.

We may regard the relation, Eq. (5.4), from an alternate point of view: the right side of Eq. (5.4) is, by definition, a *two-dimensional convolution* of $E_x(y', z')$ with $\exp ikf(y'^2 + z'^2)$, where $f \equiv [2(x-x')]^{-1}$:

> *Propagation in free space a distance L (under the above assumptions) performs a convolution of the initial scalar field with an impulse response* $\exp ik(y^2 + z^2)/2L$.†

But, recalling the discussion of Sec. 4.2:

> The elementary azimuth compression operation to be performed on the SAR data is realized by creating a wavefront at a plane whose two-dimensional scalar field is the received data and *allowing it to propagate an appropriate distance L through free space.*

The modifier "elementary" is used because we recall that the quadratic phase—and hence the appropriate distance L—depends on the range of the scattering object.

† Of course, the Kirchhoff–Huygens diffraction integral itself, Eq. (5.1), is similarly a convolution, but with a more complicated "impulse response." This, in fact, is a consequence of the linearity and constant coefficients of the scalar wave equation.

It will be convenient later to note now that as $|x - x'| \to 0$, we must have $E_{x'}(y, z) \to E_x(y, z)$:

$$\lim_{|x-x'|\to 0} \left\{ \frac{\exp\left(\dfrac{ik(y^2 + z^2)}{2|x - x'|}\right)}{|x - x'|} \right\} = \delta(x, z), \tag{5.5}$$

where $\delta(y, z)$ is the two-dimensional Dirac generalized function.

Fraunhofer zone. Generally, the calculation of the integral in Eq. (5.4) is accomplished only numerically, sometimes, depending on the nature of $E_{x'}(y', z')$, with the help of tabulations of "Fresnel integrals." We note, however, that if in addition to the above assumptions concerning small scattering/propagating angles, we have the approximation

$$\max_{(y', z') \in A'} \left\{ k \frac{y'^2 + z'^2}{2R} \right\} \ll 1, \tag{5.6}$$

that is, if $a'/\lambda R \ll 1$, where a' is the area of the aperture A', then we have, setting $|x - x'| \equiv R$,

$$E_x(y, z) \approx \frac{-ik}{4\pi} \frac{\exp(ikR)}{R} \exp\left(ik \frac{y^2 + z^2}{R}\right)$$
$$\cdot \iint_{A'} \exp\left(i \frac{ky}{R} y' - i \frac{kz}{R} z'\right) E_{x'}(y', z')\, dy'\, dz'. \tag{5.7}$$

The region where $R \gg a'/\lambda$ is called the *Fraunhofer, or far-field, zone*; Equation (5.7) therefore states that

in the Fraunhofer zone, the scalar field is given, within a quadratic phase, by the (scaled) Fourier transform of the aperture scalar field.

The fact that a coherent optical system can compute a Fourier transform implies that it can realize any LIV system—assuming we can create specified distributions at specific locations, as of course we can (Sec. 5.1.3). But the "far-field" for a 35 mm aperture requires that $R \gg a'/\lambda = 10\,\text{km}$!! We therefore need to "bring" this plane into a practical distance: we are not surprised that this can be done by a lens—one with a quadratic phase which essentially can simulate a large amount of "free space."

5.1.2 Operations with a Quadratic Lens

The most important element of an optical system is a "spherical lens." For our purposes a "spherical lens" is defined by the relation

$$E_o(y, z) = \exp ikf(y^2 + z^2)\, E_i(y, z), \tag{5.8}$$

where $E_i(y, z)$ is the field impinging on the plane of the lens and $E_o(y, z)$ the emerging field at the plane of the lens; exp $ikf(y^2 + z^2)$ is called the "transmission function" of the lens. This relation, in fact, is a good approximation to the operation performed by a real lens—as must be true, or else such a lens is of no interest to us here. A *cylindrical lens* operates in only one dimension; e.g.,

$$E_o(y, z) = \exp ikfy^2 \, E_i(y, z). \tag{5.9}$$

We shall refer to both spherical and cylindrical lenses as *quadratic lens*. These optical elements are important because (1) their operation on a scalar field can be a Fourier transform—and hence integration, and (2) they are among the very few optical elements that exist for use in an optical system.

In order to describe the operation of a quadratic lens sufficiently well for our purposes, consider the optical system of Fig. 5.2. We define

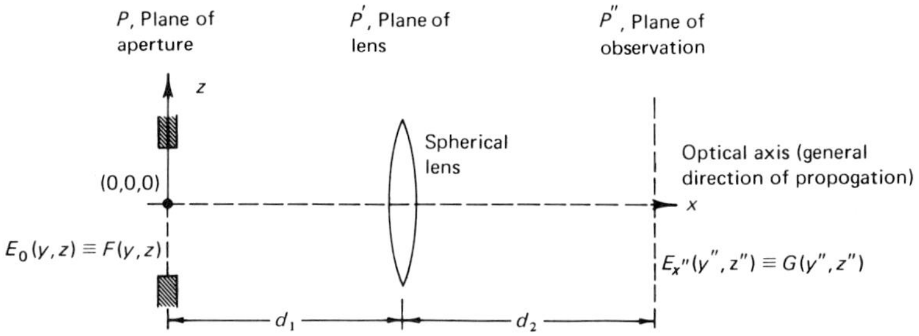

FIG. 5.2 Geometry for calculation of operation of a quadratic lens.

$$\delta_i \equiv 1/d_i, \qquad i = 1, 2. \tag{5.10a}$$

$$\theta(y, z) \equiv \tfrac{1}{2}(y^2 + z^2), \tag{5.10b}$$

and denote the transmission function of the lens by

$$\exp ikf\theta(y', z'),\dagger \tag{5.10c}$$

Using (twice) expression (5.4) characterizing diffraction (i.e., propagation) through free space, and expressions (5.8) and (5.10c) for the operation of the quadratic lens, we find the scalar field $G(y'', z'')$ at the plane of observation to be, within a constant complex scale factor,

$$G(y'', z'') = \iint \exp[ik\,\delta_2\,\theta(y'' - y', z'' - z')] \cdot \exp(ikf\theta(y', z') \cdot \iint_A \exp[ik\,\delta_1\theta(y' - y, z' - z)]F(y, z)\,dy\,dz \cdot dy'\,dz'. \tag{5.11}$$

† We have already noted that very small scattering angles are of interest in these systems; it is therefore reasonable to assume expression (5.10c) holds over all of P' of interest—that is, we ignore any finite lens "entrance" aperture.

The integration over the P' plane can be done using the generalized Fourier transform relation

$$\int dx \exp(-iux)\exp(i\alpha x^2) = c\exp(-iu^2/4\alpha) \tag{5.12}$$

where c is a complex constant. Thus, within such a complex constant,

$$G(y'', z'') = \exp\{\tfrac{1}{2}[ik\,\delta_2(y''^2 + z''^2)]\} \iint dy\, dz\, F(y, z) \exp\{\tfrac{1}{2}[ik\,\delta_1(y^2 + z^2)]\}$$

$$\cdot \exp\left\{-i\frac{[k(\delta_2 y'' + \delta_1 y)]^2}{2k(\delta_1 + \delta_2 + f)} - i\frac{[k(\delta_2 z'' + \delta_1 z)]^2}{2k(\delta_1 + \delta_2 + f)}\right\}. \tag{5.13}$$

Equation (5.13) is the basic description of the operation of a quadratic lens.

Imaging. By an *image*, G, *of* F we mean that G is a replica, within a scale or magnification factor, of F. In order that the configuration of Fig. 5.2 produce a field G at P'' that is an image of F at P, the integrand appearing in Eq. (5.13) must be essentially the product of F by a delta function. But by Eq.(5.5) we must, therefore, have

$$\delta_1 + \delta_2 + f = 0. \tag{5.14}$$

Equation (5.13) now becomes

$$G(y'', z'') = \exp[\tfrac{1}{2}ik\,\delta_2(y''^2 + z''^2)] \iint du\, dv\, F\left(\frac{u}{\delta_1}, \frac{v}{\delta_1}\right)$$

$$\cdot \exp\left[\frac{ik(u^2 + v^2)}{2\delta_1}\right] \cdot \delta(\delta_2 y'' + u, \delta_2 z'' + v) \tag{5.15}$$

$$= \exp[\tfrac{1}{2}ik(\delta_2 + \delta_2{}^2/\delta_1)(y''^2 + z''^2)]F\left(-\frac{\delta_2}{\delta_1}y'', -\frac{\delta_2}{\delta_1}z''\right).$$

Thus, if Eq. (5.14) is satisfied, Eq. (5.13) reduces to Eq. (5.15), which is the imaging operation performed by a quadratic lens: within a (spatially varying) quadratic phase, G is a reversed image of F, scaled by the *magnification*

$$M \equiv \frac{\delta_1}{\delta_2} = \frac{d_2}{d_1}. \tag{5.16}$$

The spatially varying quadratic phase $\exp[\tfrac{1}{2}ik(\delta_2 + \delta_2{}^2/\delta_1)(y''^2 + z''^2)]$ is not important *if*, e.g., the field G is recorded on film which responds only to the magnitude of the impinging field. But if, e.g., subsequent general operations are to be performed on the propagating field, then the phase must be accounted for.

Example 1. Suppose that $d_1 \to +\infty$, $\delta_1 \to 0$. Then by Eq. (5.15), the wavefront is concentrated, or *focused*, to a very small spatial extent at $d_2 = -1/f$: then P'', located at $d_2 = -1/f$, is called the *back focal plane* of the quadratic lens. In this limiting case, one would have to account for the finite size of the lens aperture in order to find the structure of the actual image; in fact, imperfections in any real lens would likely determine the behavior near focus.

Example 2. Suppose that $d_2 \to \infty$, $\delta_2 \to 0$. Then by Eq. (5.15), the image of an object at $d_1 = -1/f$ would appear at $x'' = +\infty$. The plane P is then called the *front focal plane* of the quadratic lens.

Example 3. Suppose that $d_1 = d_2 = -2/f$, $\delta_1 = \delta_2 = -f/2$. Then the image has unit magnification.

The Fourier transform operation. In order to have G be the Fourier transform of F, we must eliminate all phases in y and z except a phase of the form $\alpha y''y + \alpha z''z$. Expanding the quadratics appearing in Eq. (5.13), we see that the quadratic phase involving (y, z) is eliminated, if and only if

$$k\delta_1 = \frac{k{\delta_1}^2}{\delta_1 + \delta_2 + f},$$

i.e., if and only if

$$\delta_2 = -f \tag{5.17}$$

independently of δ_1. Therefore, *except for a spatially varying quadratic phase, the distance of P from P′ does not affect the Fourier transform operation.* For example, if we were going to record the Fourier transform on photographic film, the distance d_1 of F from the quadratic lens does not matter. Also, e.g., if we wish to utilize the Fourier transform only at $(y'', z'') = (0, 0)$, the value of d_1 does not matter: we shall use this fact in Sec. 5.2.

The quadratic phase involving (y'', z'') is eliminated, if and only if

$$k\delta_2 = \frac{k{\delta_2}^2}{\delta_1 + \delta_2 + f}$$

i.e., if and only if

$$\delta_1 = -f, \tag{5.18}$$

independently of δ_2. With the conditions, Eqs. (5.17) and (5.18), the coefficients of $y''y$ and $z''z$ both equal $-kf$, and hence Eq. (5.13) becomes

$$\begin{aligned} G(y'', z'') &= \iint dy\, dz \exp[i(kfy'')y + i(kfz'')z]F(y, z) \\ &= \tilde{F}(-kfy'', -kfz''). \end{aligned} \tag{5.19}$$

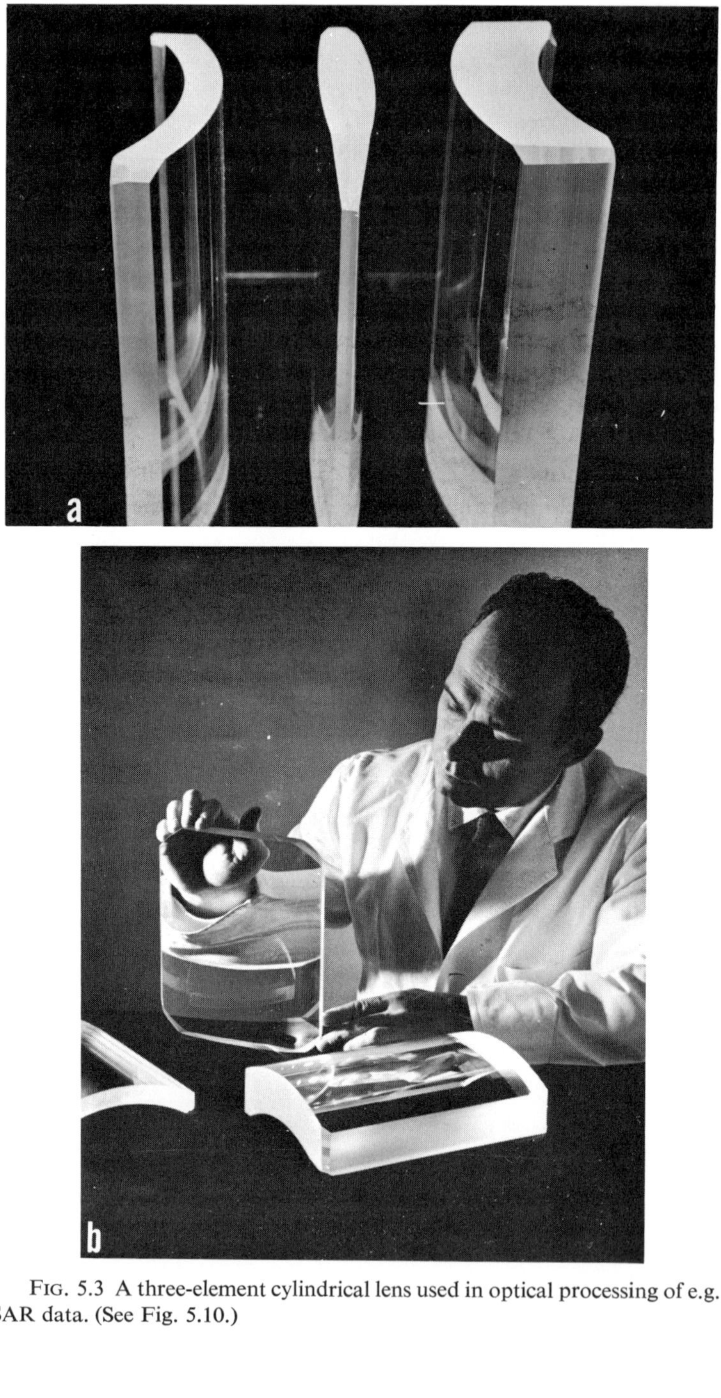

FIG. 5.3 A three-element cylindrical lens used in optical processing of e.g., SAR data. (See Fig. 5.10.)

G is a scaled version of the Fourier transform $\tilde{F}$ of F.

Integration. We note from this last result, Eq. (5.19), that

$$G(0, 0) = \iint dy\, dz\, F(y, z). \tag{5.20}$$

But, as was noted implicitly above [in the discussion just after Eq. (5.17)], if we examine the field at $(y'', z'') = (0, 0)$, *the plane P may be any arbitrary distance d_1 from the quadratic lens plane P′ without changing the value of the integral in Eq.* (5.20). The important point is that an optical system employing a quadratic lens can perform the operation of integration.

Anamorphic systems. We have carried the discussion of this section through for a spherical lens. But every expression is free of any cross products of y's and z's: that is, every expression factors into a y-description and a z-description. Thus, the development for a cylindrical lens exp $ikfz'^2$, goes through in an obvious way, giving an optical system that acts differently in the y and z orthogonal transverse coordinates and is therefore called "anamorphic." We can have imaging, Fourier transforming, or integrating in the z-direction, and simply free space diffraction in the y-direction.

We, of course, need anamorphic optical systems for the SAR system. A cylindrical lens used in optical processsors is shown in Fig. 5.3.

5.1.3. Transmission Functions

The fact that optical systems can perform the operations of imaging, Fourier transforming, and integrating is of little practical consequence unless there exists a method of creating specified scalar fields at specified locations in an optical system. In fact, this is commonly done by use of "transmission functions." That is, it is possible to make *transparencies* that, if placed at a plane P with an impinging scalar field $E_i(y, z)$, the emerging scalar field $E_o(y, z)$ is given by the simple *multiplicative* relation

$$E_o(y, z) = H(y, z)E_i(y, z). \tag{5.21}$$

We, of course, are using the complex representation of the scalar fields, the actual emerging field being given by $\text{Re}\{E_x(y, z)e^{i\omega_o t}\}$. The multiplicative complex factor $H(y, z)$ is called the *transmission function* and, for our purposes, defines the transparency. The magnitude $|H|$ of H is assumed to be bounded by 0 and 1,

$$0 \leq |H| \leq 1,$$

so the transparency is "passive"—that is, the energy of the emerging field cannot exceed the energy of the impinging field.

In order that an optical device be characterized a simple multiplicative operation, there must be no diffraction within the optical device: in this sense, it must be optically "thin."

Example 1. We have already used the transmission function with uniform amplitude and quadratic phase to approximate the optical operation performed by spherical and cylindrical lenses.

Example 2. An ordinary photographic film negative transmission function has approximately only a magnitude, any phase variation being due to imperfections—varying base thickness, for example.

We shall continue the discussion of transmission functions, assuming the transparency is produced by use of ordinary photographic film. But the discussion applies to any medium or method used to produce transparencies that have the following property:

The medium responds to impinging energy and is passive.

The discussion here concerns the practical consequences of this property. Of course, any recording medium has a finite spatial bandwidth, as discussed in Sec. 3.5, a nonlinear response, as will be discussed in Chapter VII, and contributes errors, as was discussed in Chapter IV, and will be discussed in Chapter VI.

Thus, if $E_i(y, z)$ is the scalar field impinging on the (unexposed) medium, the transmission function will be approximately

$$H(y, z) = C \int_T |E_i(y, z, t)|^\alpha \, dt, \tag{5.22a}$$

where $C > 0$ and $\alpha > 0$ are constants depending on the nature of the medium (and its development in the case of photographic film), and T is the time duration of the exposure. For our purposes $|E_i(y, z, t)|$ does not depend on t, and for present purposes it is sufficient—and considerably simpler—to take $\alpha = 2$. Now

$$H(y, z) = |E_i(y, z)|^2 \tag{5.22b}$$

within an uninteresting multiplicative constant.

As we have already noted in Sec. 1.4, the model, Eqs. (5.22a) and (5.22b), has the consequence that we cannot simply directly expose the film (e.g.) to the desired scalar field. Rather we must use a trick such as the hologram method to get the desired field recorded on a bias and a spatial carrier, or "offset" frequency. Of course, this presupposes that, in fact, we can create at a specified plane the requisite scalar field: this may be very difficult. Another possibility is to precompute the desired field and to program a narrow light beam to expose the film while scanning over it, incorporating the offset frequency much as it is in the SAR system. Fortunately, complicated phases are not normally required of transmission functions in the SAR system, free space diffraction providing the essential phase.

Assuming that we have managed to construct a transparency whose transmission function is

$$H(y, z) = |H(y, z)| = b_o + |F_{y,z}| \cos(\omega_o y + \phi_{y,z}), \tag{5.23a}$$

we can employ the optical system of Fig. 5.4 to recover the desired scalar

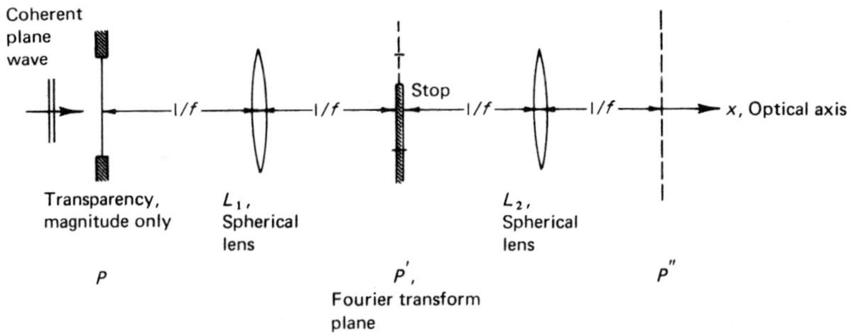

FIG. 5.4 Coherent optical system to produce a complex (i.e., phase and amplitude) scalar field from a real (i.e., amplitude only) transmission function.

field $F_{y,z} \equiv |F_{y,z}| \exp i\phi_{y,z}$. If we insert the transparency in an aperture at plane P and illuminate it from the left with a coherent plane wave, propagating along the optical (x) axis, the scalar field emerging from P is given by Eq. (5.23a), which can be rewritten as

$$H(y, z) = b_o + \tfrac{1}{2}F_{y,z}e^{i\omega_o y} + \tfrac{1}{2}\bar{F}_{y,z}e^{-i\omega_o y}. \tag{5.23b}$$

The spherical lens L_1 produces the Fourier transform $\tilde{H}$ of H at plane P': the first term, b_o, of H produces a scalar field confined to a small location about the optical axis, the second about a location corresponding to the offset ω_o, and the third about a location corresponding to the offset $-\omega_o$. If ω_o is chosen sufficiently large, these three terms are separated, and we may physically prevent any of them from propagating further down the optical system.

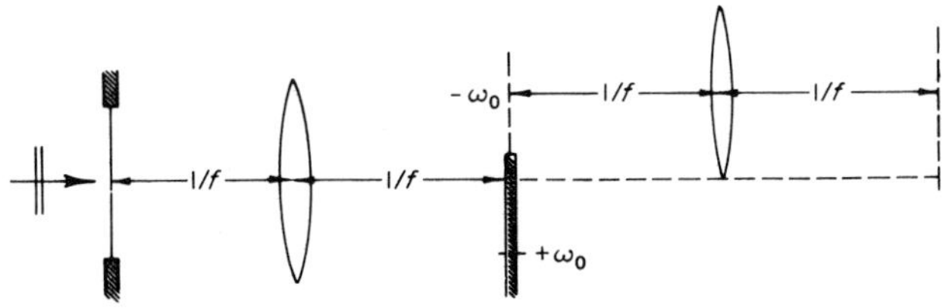

FIG. 5.5 Coherent optical system to produce a complex scalar field (without spatial offset frequency) from a real transmission function.

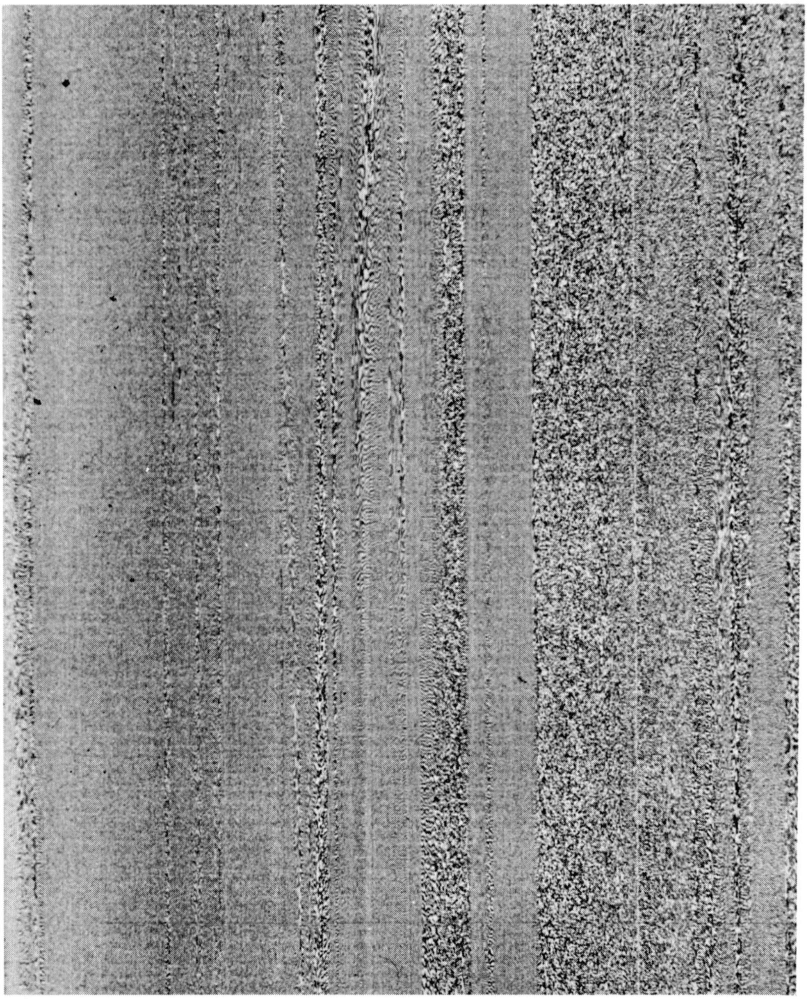

FIG. 5.6 An example of SAR signal, or data, film.

In particular we allow only the spectrum, $\tilde{F}(\omega - \omega_o)$, due to $Fe^{i\omega_o t}$, to pass. The lens L_2 now Fourier transforms again—i.e., inverts $\tilde{F}(\omega - \omega_o)$—to give $F_{y,z}\, e^{i\omega_o t}$ at the plane P''.

It may be that $F_{y,z}$, vice $F_{y,z}\, e^{i\omega_o y}$, is actually desired. This "heterodyning" can be accomplished in the remarkably trivial way of aligning the optical axis of L_2, not with that of L_1, but laterally to the position corresponding to ω_o. For then the spectrum $\tilde{F}(\omega - \omega_o)$ appears to L_2 simply as $\tilde{F}(\omega)$, and hence produces at P'' simply F. (See Fig. 5.5).

A typical sample of SAR signal film is shown in Fig. 5.6.

5.2. OPTICAL COMPRESSION

5.2.1. Spatial Domain Processor

We recall from the discussion of Chapter IV, especially Sec. 4.2, that a minimal processing of the data is accomplished by a LIV filter whose impulse response has uniform amplitude and phase $-\frac{1}{2}ik_a x^2$. We have just observed (Sec. 5.1.1) that, in a coherent optical system, such a filter is realized by free spacc.

Thus, suppose a single azimuth history was recorded on ordinary photographic film (as discussed in Sec. 5.1.3), placed in an aperture, and illuminated with a coherent plane wave propagating along the optical axis (see Fig. 5.7a).

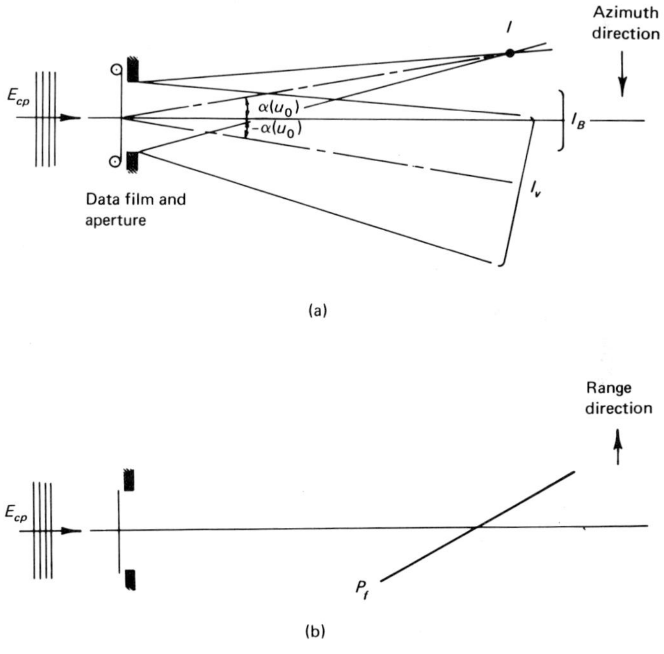

FIG. 5.7 The geometry of the reconstruction, or compression of the SAR data. (a) Slant range plane geometry showing physical separation of image (I), vertical image (I_v) and bias (I_B) wavefronts; $\alpha(u_o)$ is offset angle corresponding to offset frequency u_o. (b) Vertical plane geometry showing inclination of plane of focus, P_f. (E_{cp} is a coherent plane wave.)

Then the image will automatically be formed at a plane an appropriate distance fron the aperture plane, and at an angle from the optical axis determined by the offset frequency.†

As we have noted, k_a depends on R, the range to the object giving rise to the azimuth history. Therefore, the ensemble of azimuth histories require varying "amounts of space" to diffract to their respective images. Obviously, the images appear on a plane inclined to the optical axis, because their recorded wavefronts were diffracted from such a plane (see Fig. 5.7b). This is inconvenient: we would like to "erect" this plane so that its normal is along the optical axis. Additionally, we must account for the diffraction that has occurred in the range direction while the wave propagates to any such erect plane. The required optical system, which must therefore be anamorphic, we now describe.

In order to erect the inclined plane of the image, it is apparent that we need an optical device that acts much like a cylindrical lens (operating in the azimuth direction), except that its focal length must vary linearly as a function of range. By an apparently wild stroke of fortune, such a lens—called a "conical lens"—exists among the very limited assortment of optical devices. We can therefore align the inclined front focal plane of the conical lens with the inclined plane of the images, thereby (Sec. 5.1.2) reimaging this plane to $+\infty$.‡

In order to reimage in the range dimension to $+\infty$, we obviously can use a cylindrical lens, acting of course in the range dimension, by placing its front focal plane into coincidence with the data film.

Now we have an erect and simultaneously (in range and azimuth) focused image but must obviously "bring it in" to some conveniently located plane. This can be done with a spherical lens, reasoning as follows: if a wavefront produces an image at $+\infty$, then at any finite plane the wavefront must be, within a quadratic phase, the Fourier transform of that image; so if we intercept this wavefront with a spherical lens, then in the back focal plane of the lens must appear, within a quadratic phase, the Fourier transform of the Fourier transform of the image—that is, the image. Since we usually record

† The offset frequency has been chosen to separate the *spectra* of the three [Eq. (5.23b)] components of the field. The Fourier transform appears at a sufficiently large distance [Eq. (5.6b)] from the aperture, and its argument is an *angular* variable [Eq. (5.7)]. It can be shown that this choice of offset frequency does ensure the separation of the desired component from the others.

‡ We noted in Sec. 3.2 that the SAR data was generated by a linear, quasi-invariant mapping, and in Sec. 4.1 that the optimum processors would also be quasi-invariant: the conical lens is precisely the realization of the slowly varying behavior of the processor.

this image, or compressed signal, at this plane the quadratic phase may be ignored.

Of course we can place only a small amount of data film in the optical system's aperture at one time. We therefore record the image only over a narrow slit, extended in the range direction; the data film and the recording (image) film are now translated uniformly to process the requisite length of data.

The coherent optical processing system is shown in Fig. 5.8. This is not

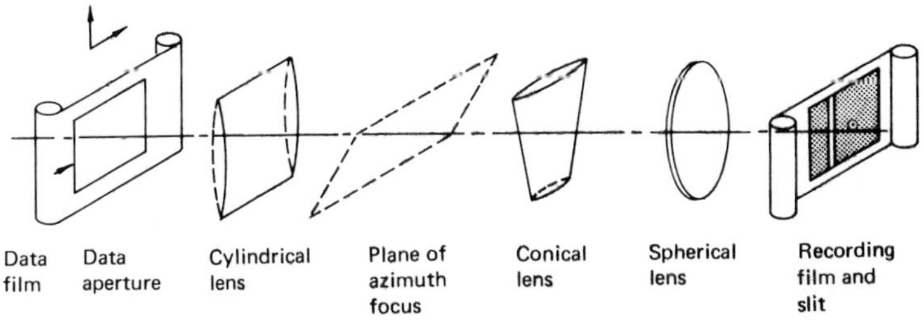

FIG. 5.8 Coherent optical system to compress, or image, the data film of the SAR system.

the most common configuration, which can be obtained by the following reasoning. The fact that an azimuth history focuses itself means that it acts like a lens located at the plane of the data film. We can now place a (conical) lens against this plane, so that the combined real and fictitious lens will focus at $+\infty$.

The final configuration is shown in Fig. 5.9.

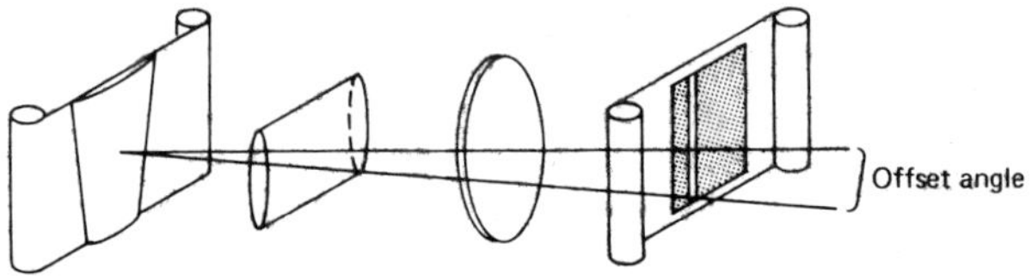

FIG. 5.9 Coherent optical system to compress/image the data film of SAR systems. (Second form.)

A physical realization of the processor of Fig. 5.9 is shown in Fig. 5.10, a very high performance and flexible optical processor at the University of Michigan.

FIG. 5.10 Physical realization of the coherent optical system of Fig. 5.7. (A) Coherent light source (Laser), (B) Collimating and cornering optics, (C) signal (data) film transport mechanism, (D) location of processor entrance aperture and conical lens, (E) cylindrical lens (see Fig. 5.3), (F) spherical lens, and (G) image recording camera at processor output slit.

5.2.2. Simultaneous Range and Azimuth Compression

As we have noted, one of the most convenient signal modulations is linear FM. If such a signal modulation is used, the ensemble of scattered signals therefore have a quadratic phase in the range (y) dimension. Therefore, so far as the range dimension is concerned, when the data film is illuminated as usual by a coherent plane wave, the emerging wavefront has a quadratic phase that is equivalent to that of a wavefront from an object at a different axial location. We, therefore, need only to place this plane coincident with the front focal plane of the range cylindrical lens, the compressed image then appearing of course in its back focal plane. (Of course, we readjust the azimuth cylindrical lends so that its back focal plane is coincident with that of the range cylindrical lens.) It turns out that this axial displacement is easily achieved in practice, so that with this trivial change the coherent optical system of Fig. 5.9 can simultaneously compress the two-dimensionally dispersed data that is recorded when a linear FM signal modulation is used in the SAR system.

5.2.3. Alternate Derivation of Optical Processor

The coherent optical processor is of such great importance to SAR systems that another explanation, or derivation, of it seems well worthwhile. Additionally, the present point of view clarifies the generalization necessary for more complicated LIV optical processors.

The minimal processing of the SAR data requires, again, a convolution of the data with an LIV filter of impulse response $\exp \frac{1}{2}ik_a x^2$ in the azimuth dimension, k_a depending weakly on range, denoted y. But we have noted that such a transmission function is realized by a conical lens: if we place this lens against the data film which is illuminated by a coherent plane wave, the emerging wavefront is the scalar field

$$Z(x - x_o, y) \exp \tfrac{1}{2}ik_a x^2. \tag{5.24}$$

The translation x_o is obtained by azimuth direction translation of the data film of course.

Now we must integrate Eq. (5.24) to compute the required convolution for the value x_o

$$G(x_o) = \int Z(x - x_o, y) \exp[\tfrac{1}{2}ik_a x^2]\, dx. \tag{5.25}$$

But we know how to perform integration with a quadratic lens (Sec. 5.1.2): we can locate a cylindrical quadratic lens, acting in the azimuth direction, such that the plane of the data film coincides with the front focal plane of the lens, and we examine the Fourier transform plane—its back focal plane —at $x = 0$.

We must, of course, account for the diffraction in the range (y) direction: if we reimage at infinity, then we can replace the azimuth cylindrical lens by a spherical lens that will reimage (in range) in its back focal plane.

Since the squared-magnitude of the image is recorded, only two distances are critical: (i) the range cylinder must have the signal film in its front focal plane; and (ii) the image film must be in the back focal plane of the spherical lens.

The convolution is computed and recorded for all values x_o by uniformly translating the signal and image films in the azimuth direction. The realized optical system is of course that of Fig. 5.9.

5.3. LIV FILTERING

As was seen in Chapter IV, especially Sec. 4.3, the optimum processing can be a LIV filter more complicated than the simple compression filter $\exp \frac{1}{2}ik_a x^2$ whose realization we have just discussed.

As a special case, if the other LIV filter has an impulse response $h(t) \exp \frac{1}{2}ik_a x^2$, where h is real, then h can be realized in an optical system by a transmission function that is an amplitude function only. Such a transmission function can be directly recorded on film and the developed film introduced into the optical system at the plane of the data film and conical lens. In particular, in this simple way we can realize the *matched filter* for the SAR system —matched, that is, to the signal due to a "point" scattering object.

5.3.1. Special Case

If the LIV filter is complex, then, assuming we are restricted to using a film-like recording medium, we must use the hologram technique (or some equivalent) to produce the transmission function and use the optical circuitry of Fig. 5.3 or 5.4 to produce, *at the plane of the data film*, the required complex function. Because the optimum filters of Chapter IV for SAR systems all embody the compression operation with the quadratic phase $\frac{1}{2}k_a x^2$, the conical lens would usually be used to assist in the synthesis of the complex impulse response.

For example, the optical system of Fig. 5.11 realizes (in principle) any LIV system with impulse response $K(x) \exp \frac{1}{2}ik_a x^2$, with K generally complex.

5.3.2. Frequency Domain Processor

To realize more general LIV processors, we make available the Fourier transform of the data in order to multiply it by a transmission function equal to the transfer function of the desired LIV filter. We, of course, can do this with a spherical lens: we place the data in its front focal plane, and the trans-

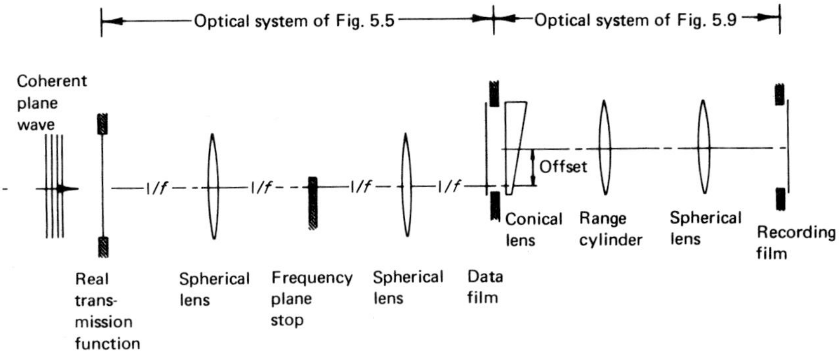

FIG. 5.11 Coherent optical system to realize an LIV processor with impulse response $K(x, y) \exp(\frac{1}{2}ik_a x^2)$, K complex.

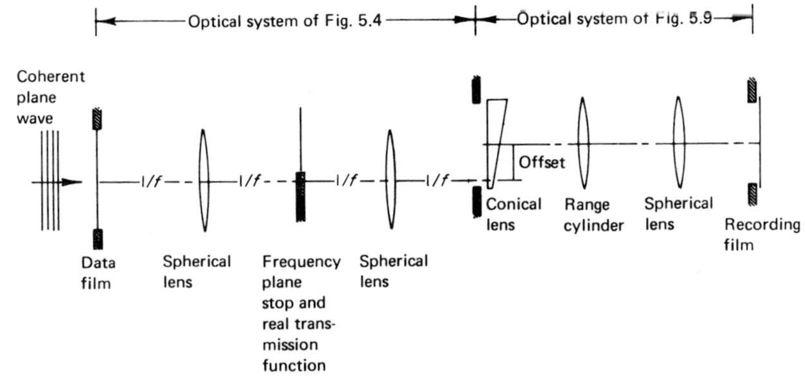

FIG. 5.12 Coherent optical system to realize a general class of LIV processors for the SAR system: the "frequency domain processor."

form appears in its back focal plane. Since the data film records a real function, it produces three component wavefronts also; only the desired one is passed at the frequency plane.

Assuming that we are using a film-like medium, we must use the hologram technique (or some equivalent) to record the *transfer function* on an offset frequency. This means that, when we use another spherical lens to Fourier transform again (i.e., invert), the three usual components of the wavefront separate spatially at the *spatial domain plane.*

Since generally—e.g., for all of those of Chapter IV—the optimum filter can be represented as a sequential operation by a linear, quasi-invariant filter with impulse response $\exp \frac{1}{2}ik_a x^2$ preceded (or followed) by a LIV filtering operation, the more general linear processors usually utilize the optical system of Fig. 5.9 as the "final" processing step, preceded by an optical circuit that first performs the remaining LIV filtering. Such a system is shown in Fig. 5.12. (Note that it is quite similar to the system of Fig. 5.11.) This type of optical processor is often called a "*frequency domain processor.*"

COMMENTS AND REFERENCES

The application of the techniques of coherent optics to data processing has undergone a tremendous development in the last decade. Some early discussions are Elias (1952), O'Neill (1956), and Cutrona *et al.* (1960).

A discussion on realizing complex (i.e., amplitude and phase) functions and filters was given by Vander Lugt (1964), who has also given a simple, formal "operational" analysis and synthesis procedure for these systems: Vander Lugt (1966).

Optical processing in the SAR context has been discussed by Cutrona *et al.* (1966), and Brown (1967). This application of the conical lens is one of the many contributions of E. N. Leith.

VI

Phase and Motion Errors

Part I: Phase Errors

6.1. INTRODUCTION

If $f(t) \equiv |F(t)| \cos[\omega_o t + \theta(t)]$ is a signal (function), then $f(t) \equiv |F(t)| \cos[\omega_o t + \theta(t) + \gamma(t)]$ is said to be the signal with a *phase error*, γ. The functions θ and γ are assumed to be real. In the complex representation, the signal with phase error γ is

$$F_\gamma(t) \equiv |F(t)| \exp[i\omega_o t + i\theta(t) + i\gamma(t)]$$
$$\equiv F(t) \exp[i\omega_o t + i\gamma(t)].$$

Sometimes, e.g., in SAR systems, some discretion is required in representing f_γ by its complex modulation F_γ, since the phase error γ may destroy the usual "narrow-band" assumption.

A perturbing influence of a similar form could be defined in the frequency domain instead of the time domain: if f is a signal (function) whose complex representation has spectrum $\tilde{F}$, then f_δ with complex representation of spectrum $\tilde{F}e^{i\delta}$ is said to be the signal with a *dispersive error*, δ.

By definition, the investigation of the time-domain effects of *dispersive* errors is the analysis of the effects of a factor exp $i\gamma$ on the Fourier transform. But it turns out that the investigation of the time-domain effects of *phase* errors on the output of a phased-matched filter of a *linear* FM signal is also equivalent to the analysis of the effects of a factor exp $i\gamma$ on the Fourier transform. Therefore, (with a change of scale), the analysis of the effects on linear FM signals will carry over directly to describe the effects of dispersive errors. We shall, therefore, delay mention of dispersive errors until Sec. 6.5.

We note, first, that phase errors are a type of *multiplicative* error—by definition of course—and are, therefore, of a different nature than the *additive* errors discussed in Chapter IV. There is no natural separation into a "signal" term and an additive "noise" term; nor, e.g., is the output of an LIV system with such an input separable into two such terms.

We note, second, that phase errors, by definition, phase modulate the signal, generally increasing the spectral spread of f_γ relative to that of f. Since the total energy of f_γ is invariant with respect to γ—i.e., the total energy of f_γ equals the total energy of f for any γ—this means that the signal-to-additive-noise ratio is less for f_γ than for f.

Next, we note that in all of the systems of Chapter IV, a constant—perhaps unknown—phase was assumed and accounted for, so that those systems are not affected by a constant phase error: only *changes* in phase error are of importance. Further, because of the "many-to-one" mapping performed by the function exp $i\gamma$, when γ exceeds 2π rad, we can expect an ambiguity between, e.g., the variance and the bandwidth of $\{\gamma(t)\}$. As a matter of fact, this is not particularly interesting here, because such large fluctuating phase errors essentially destroy system performance by any reasonable measure.

With regard to the MSE criterion, we will not be surprised to find the MSE increased by the presence of phase errors. Some care is required in using this criterion. For example, a linear phase error $\gamma_1 t$ will, as we have seen, merely displace the output image of a linear FM signal when matched filtered (assuming it remains within the passband of the filter): in many applications, such an effect, even if of unknown magnitude, is of no importance. As a further example, a known quadratic phase error can be removed and, if unknown, can be estimated and partially removed (Chapter X) in a trivial manner by an optical processor (Chapter V). Therefore, we shall employ the MSE criterion only in conjunction with broad bandwidth (relative to $2\pi/X$) phase errors (Sec. 6.3).

The point of view in this chapter is to calculate the effects of phase errors on the systems of Chapter IV: those optimum under various criteria accounting for *additive* noise. In Chapter X, we shall reconsider the optimum system design problem by formulating criteria that a priori account for the presence of phase errors. We saw in Chapter IV that the various optimum processors were very similar, all being linear and quasi-invariant filters that included a quadratic phase and had, generally, a fairly uniform amplitude. This restriction considerably enhances the value of numerical analysis, and we shall, therefore, use this method extensively.

The systems resulting from the resolution, $\mathrm{SNR}_{\max}$, and ML criteria of Chapter IV are naturally analyzed by dividing the phase errors into two classes:

1. Small, small-bandwidth phase errors that enable γ to be approximated by a few terms of its Taylor series.
2. Small, large-bandwidth phase errors.

One of the purposes of the present discussion is, of course, to derive bounds on tolerable phase errors for SAR system design (Sec. 6.6). We shall find that the SAR system is so sensitive to fluctuating phase errors, that only phase errors much less than a radian are tolerable—hence, the case of small phase errors is of sufficient interest here.

Sources of phase errors. There are several significant sources of phase errors in SAR systems:

1. Spurious relative motion of vehicle and scattering objects.
2. Oscillator and other system phase instabilities.
3. Imperfections in optical processor such as imperfect optical elements, film thickness variations, and imperfect light sources.
4. Atmospheric propagation.

The first is the most important and most difficult to control, especially for vehicles traveling in the lower atmosphere.

Generally speaking, phase errors due to those and other sources, produce their effect in the azimuth channel of well-designed SAR systems. We shall, therefore, carry out the discussion in the notation of the azimuth channel. Much of the discussion applies in an obvious way to the range channel, of course. In a well-designed system, the range channel is more likely to be affected by dispersive errors (Sec. 6.5).

If one examines the western (left half) region of the image shown in the frontispiece of the book, the effect of phase errors can be noted: the images of "points" are "smeared" in the east-west (along-track) direction in that region. The most likely cause was air turbulence buffeting the vehicle carrying the SAR radar set.

6.2. EFFECTS ON PHASE-MATCHED SYSTEM

From the discussions of Chapters III and IV, we see that, if $B_\gamma(x) \equiv B(x) \exp i\gamma(x)$ is the azimuth channel signal, and $\bar{B}(-x)$ is the matched-filter impulse response, then the azimuth channel output for a "point" scattering object is

$$\begin{aligned} q_\gamma(x) &\equiv B_\gamma(x) * \bar{B}(-x) \\ &= \exp(-\tfrac{1}{2}ikx^2) \int |B(y-x)| \cdot |B(y)| \exp[-ikxy + i\gamma(y)]\, dy. \qquad (6.1) \end{aligned}$$

In Eq. (6.1), we have assumed the modulation B_γ to have been approximately adequately sampled; the situation where B_γ—in fact, B—is not adequately sampled, will be discussed in Sec. 6.4. Since we are interested in $|q|^2$ (see Sec. 4.3.3), we ignore the quadratic phase factor below.

We are interested in q for x in the order of the azimuth resolution distance I_h: since $|B(x)|$ varies slowly over $X \gg I_h$ we may approximate $|B(y-x)| \cdot |B(y)|$ by $|B(y)|^2$. Thus we have

$$|q_\gamma(x)|^2 \approx \left| \int |B(y)|^2 \, e^{i\gamma(y)} \cdot e^{-ikxy} \, dy \right|^2, \tag{6.2}$$

and, therefore, the mathematical problem presented by phase errors is calculating the effects on a Fourier transform.

If the filter impulse response is matched to the signal in phase only and has, e.g., uniform amplitude, then $|q_\gamma(x)|^2$ is given by Eq. (6.2) except $|B(y)|^2$ is replaced with $|B(y)|$. Since, in the present case, $|B|^2$ and $|B|$ do not differ greatly for $x \in X$, we may expect the effects of phase error to be much the same for either system. We proceed assuming matched filtering.

Since the ensemble $\{|B(x)|^2\}$ of interest was seen to be a very restricted class of functions (see Chapters IV and VIII), numerical analysis of Eq. (6.2) is very attractive. Some of our principal results will be obtained this way.

6.2.1. Small Bandwidth Phase Errors

When the phase error γ fluctuates sufficiently slowly over the duration (synthetic aperture) X of the azimuth modulation B, a Taylor series expansion of γ is useful:

$$\gamma(x) \approx \gamma_0 + \gamma_1 x + \gamma_2 x^2 + \ldots, \qquad x \in X. \tag{6.3}$$

The constant γ_0 we have already pointed out is of no importance.

These coefficients may often be usefully considered to be changing slowly with respect to X. For example, the change of latitude with an orbital vehicle generally leads to a slowly changing γ_1: this must be accounted for in a well-designed system (Chapter X).

Linear phase errors. The linear term $\gamma_1 x$ gives, of course, a factor $\exp i\gamma_1 x$ in the integrand of Eq. (6.2): therefore, this causes a displacement (of distance γ_1/k) in the image $|q_\gamma(x)|^2$ relative to the situation where $\gamma_1 \equiv 0$. In themselves, such displacements are not usually of interest: only relative spatial locations within the overall image are usually of interest. However, the linear term $\gamma_1 x$ can cause difficulty in another way: the signal spectrum, $\tilde{B}_\gamma$, is itself displaced from its location when $\gamma_1 \equiv 0$; this can cause part or all of the spectrum—or the desired alias—to fall outside the passband of the processing filter. Such effects have been described in Sec. 2.7 and are to be avoided. It turns out that, in practice, γ_1 is often unknown and of the same order as the bandwidth of B: we shall discuss the problem of estimating and compensating for γ_1 in Chapter X.

Quadratic phase errors. The most familiar phase error is, of course, due to the quadratic $\gamma_2 x^2$. Typical results are shown in Fig. 6.1, where we have assumed that $\gamma_n \equiv 0$, $n \neq 2$, and $|B(y)|^2$ to be a prolate spheroidal wave function with $c \approx 3$—perhaps the most typical $|B(y)|^2$, in other words. From

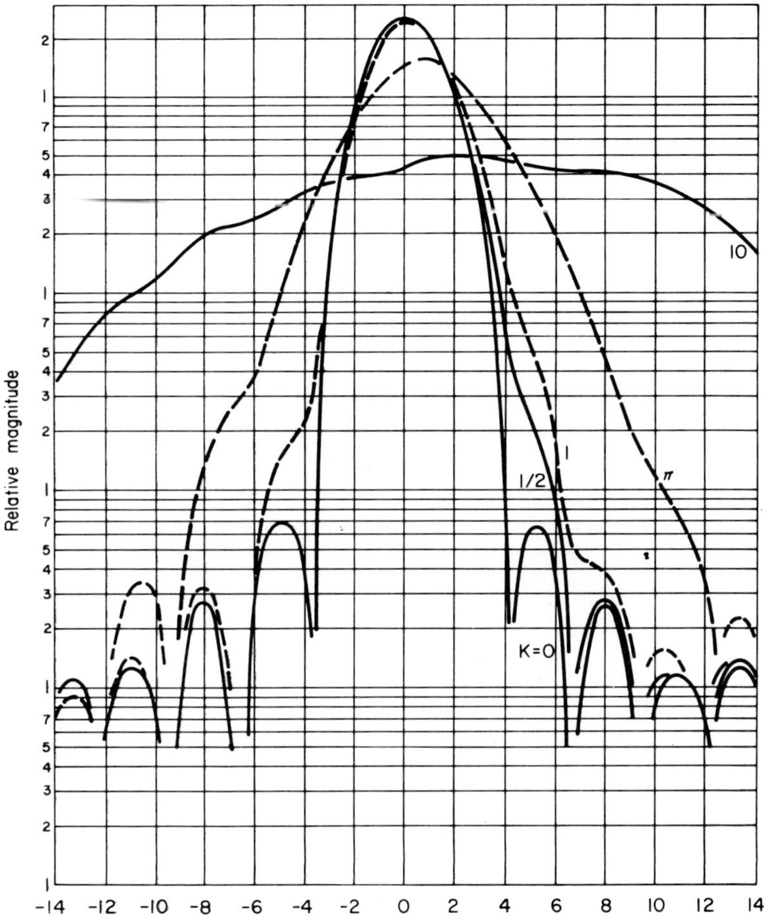

FIG. 6.1 Effect of quadratic phase errors on azimuth image. K is the phase error (in radians) at edge of aperture.

such curves, one can read or compute the dependence of any particular resolution property of interest on the amount of quadratic phase error; several of these are shown in Fig 6.2. The abscissa is $\gamma_{2m} \equiv \gamma_2(\frac{1}{2}X)^2$, the maximum value of the quadratic phase.

For example, it is apparent from Fig. 6.2a, that quadratic phase errors have a marked effect on the maximum of the compressed image and, therefore,

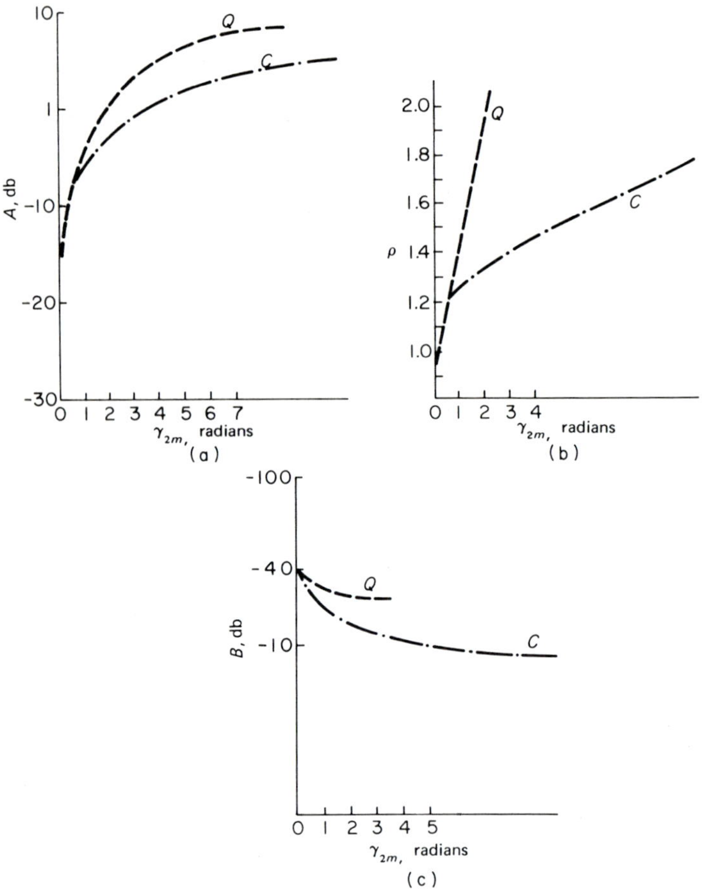

FIG. 6.2 The effects of quadratic (curves Q) and cubic (curves C) phase errors on resolution parameters. Tapered spectrum (PSWF, $c \approx 3$). (a) reduction in maximum A, relative to error-free maximum; (b) resolution ρ between 3 db points of main lobe; (c) sidelobe level B, relative to error-free maximum. All as a function of the phase error at aperture extremity.

upon the signal-to-additive-noise ratio. In other words, quadratic phase errors partially nullify transmitted power: 10 rad of quadratic phase error costs nearly 10 db in signal-to-additive-noise ratio. The reason why the curve of Fig. 6.2a "saturates" is that the energy of $|q(x)|^2$ remains approximately constant and must be distributed over a distance $2X$; the curve of Fig. 6.2c saturates for the same reason.

As a further example, Fig. 6.2c shows that quadratic phase errors "waste" horizontal aperture dimension which may have been added to obtain a certain low level of azimuth sidelobes.

From Fig. 6.2, we may state a rough "rule-of-thumb": *quadratic phase errors should be kept sufficiently small that at least* $\gamma_{2m} < 1$ rad *and preferably* $\gamma_{2m} < \frac{1}{10}$ rad.

Cubic phase errors. Again for a $|B(x)|^2$ corresponding to a prolate spheroidal wave function with $c \approx 3$ and now $\gamma_n \equiv 0, \quad n \neq 3$, the effect of a cubic phase error is shown in Fig. 6.3. The resolution parameters, A, B, and ρ, are plotted in Fig. 6.2 from Fig. 6.3; the abscissa is $\gamma_{3m} \equiv \gamma_3(\frac{1}{2}X)^3$. The general displacement of the peak response is predictable, as the "best linear fit" to the cubic is not zero as it is for the quadratic. Other general remarks are obvious.

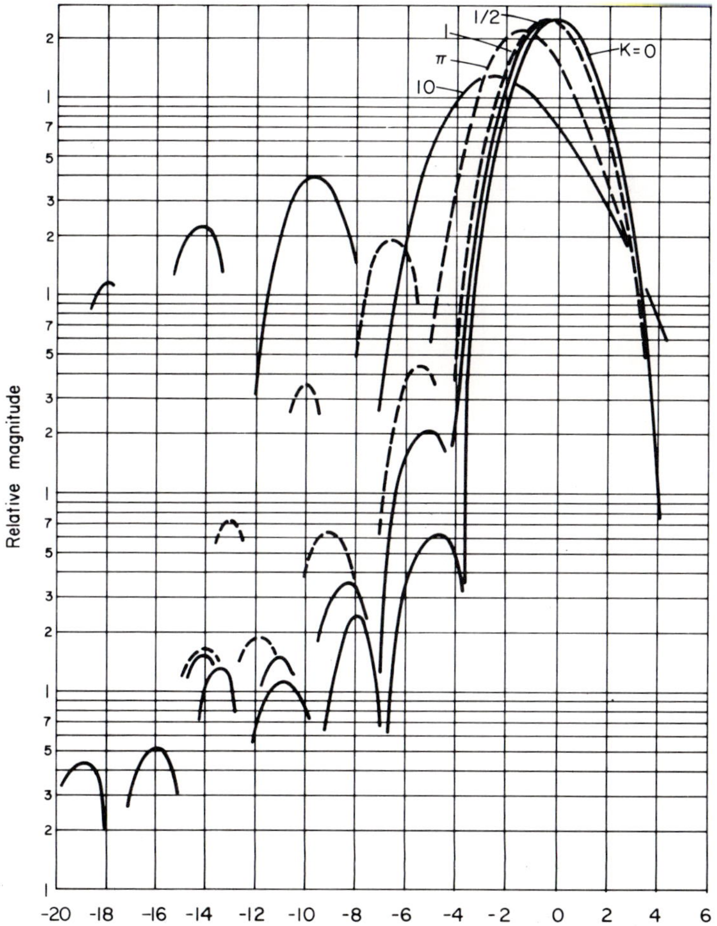

FIG. 6.3 Effect of cubic phase errors on azimuth image. K is the phase error (in radians) at edge of aperture.

The rough rule-of-thumb for cubic phase errors is the same as that for quadratic errors: *the maximum cubic phase error* $\gamma_{3m} \equiv \gamma_3(\frac{1}{2}X)^3$ *should be kept less than* 1 rad *and preferably less than* $\frac{1}{10}$ rad.

Random $\{\gamma_n\}$. One could assume that the $\{\gamma_n\}$ are random variables and compute averages of, e.g., A, B, and ρ, but this does not seem particularly illuminating. Roughly, we say that the standard deviations of γ_2 and γ_3 should be kept less than $\frac{1}{3}$ rad and preferably less than 1/30 rad as a rule of thumb.†

Analytic discussion. Since only small phase errors ($\gamma_{\max} < 1$ rad) are tolerable, one can approximate $\exp i\gamma$ by $1 + i\gamma - \frac{1}{2}\gamma^2$ and ease the problem of analyzing—and interpreting the analysis of—phase errors. We shall use this approximation below to discuss broad bandwidth phase errors (Sec. 6.2.2). But here, as an example, we calculate the effect on resolution, as measured by energy $E_{I\gamma}$ in the resolution interval I in the presence of phase errors γ, particularly quadratic phase errors.

From Eq. (6.2) we find

$$E_{I\gamma} \equiv \int |q_\gamma(x)|^2\, dx$$

$$\approx \int |B(y)|^2\, dy \int |B(\zeta)|^2\, d\zeta\, \frac{\sin \frac{1}{2}kI(y-\zeta)}{k(y-\zeta)} \exp[i(\gamma_y - \gamma_\zeta)], \qquad (6.4)$$

where we assume

$$\exp[i(\gamma_y - \gamma_\zeta)] \approx 1 + i(\gamma_y - \gamma_\zeta) - \tfrac{1}{2}(\gamma_y - \gamma_\zeta)^2.$$

By the symmetry of the quadratic form, Eq. (6.4), we can ignore the term $i(\gamma_y - \gamma_\zeta)$ in the expansion. Further, let $\gamma_y \approx \gamma_2 y^2$, a quadratic phase error; then

$$E_{I\gamma} \approx E_{I0} - \gamma_2{}^2 \int |B_y|^2 y^4 \int |B_\zeta|^2 \frac{\sin(\cdots)}{(\cdots)} + \gamma_2{}^2 \int |B_y|^2 y^2 \int |B_\zeta|^2 \zeta^2 \frac{\sin(\cdots)}{(\cdots)},$$

where E_{I0} is the energy in I when $\gamma \equiv 0$.

One can argue (Sec. 6.2.2.) that over the square $X \times X$ the integrand factor $|B_y|^2\, |B_\zeta|^2 \sin(\cdots)/(\cdots)$ is roughly constant. This allows easy integration and one finds

$$E_{I\gamma} \approx E_{I0}\left(1 - \frac{\gamma_{2m}^2}{10}\right), \qquad \gamma_{2m} < 1 \text{ rad}, \qquad (6.5)$$

where $\gamma_{2m} \equiv \gamma_2(\frac{1}{2}X)^2$ is the maximum quadratic phase. Therefore, *the resolution (measured by percent energy in the desired resolution interval) decreases*

† Chebychev's inequality guarantees that then the probability of γ_2 or γ_3 being larger than 1 rad (or $\frac{1}{10}$ rad) is quite small.

as the square of the quadratic phase error, being reduced about 10% for one radian of maximum phase error.

This argument generalizes in an obvious way but, as always with low bandwidth phase errors, some care must be taken. For example cubic errors were seen (Fig. 6.3) to cause a displacement that would effect $E_{I\zeta}$ more seriously than image quality for most purposes.

6.2.2. Large Bandwidth Phase Errors

The sources of phase errors in most systems, in particular the SAR system, are generally not known very well and are often best modeled as random phenomena. For example, vehicle motions can be due to atmospheric turbulence, which can only be usefully modeled as a random field. Therefore, we shall now assume that $\gamma(x)$ is a sample function of a random process $\{\gamma(x)\}$.

Again, because of the restricted family $\{|B|^2\}$ of interest to us here, numerical analysis is a very strong technique. One must generate (e.g., on a digital computer) the random sequence $\{\gamma(x_n),\ x_n \in X\}$ for a sufficiently fine "net," with the desired joint distributions, calculate $|q_\gamma(x)|^2$ by approximating the integral of Eq. (6.2), and note the (numerical) properties of interest. This process is repeated, and the observed properties averaged until one is satisfied, in some way, that a "good" average has been obtained.

The results of such an analysis are shown in Fig. 6.4. As in Figs. 6.1–6.3, $|B|^2$ was chosen to be a prolate spheroidal wave function with $c \approx 3$. It was assumed that $\{\gamma(x),\ x \in X\}$ was a stationary Gaussian process with mean zero, variance σ^2 and correlation function $\exp(-x^2/2F)^2$ where F is normalized by X. One hazards that the important features of the results do not depend critically upon the choice of correlation function. The averaged resolution parameters A, ρ, B, D, are all normalized with respect to their values in the absence of phase errors. (The "95% confidence intervals" are shown on the plots.)

Note that results are included for relatively narrow bandwidth phase errors—the situation when $F = 3$—that is, the correlation distance of the phase error process is three times larger than the aperture length X.

The sample functions $|q_\gamma(x)|^2$ observed when σ is large (say, $5 < \sigma < 10$) and $F \geq 0.3$ are generally very much unlike $|q_0(x)|^2$, the system response in the absence of phase errors. In fact, the resolution parameters—e.g., A, ρ, B, and D—become meaningless indicators of system performance.

The results of Fig. 6.4 cannot be summarized into as simple a "thumb-rule" as we obtained above for narrow bandwidth phase errors. But we can state the following rough guide: *if the correlation distance of random phase errors is less than the aperture length X, then their variance should be kept very much less than* 1 rad.

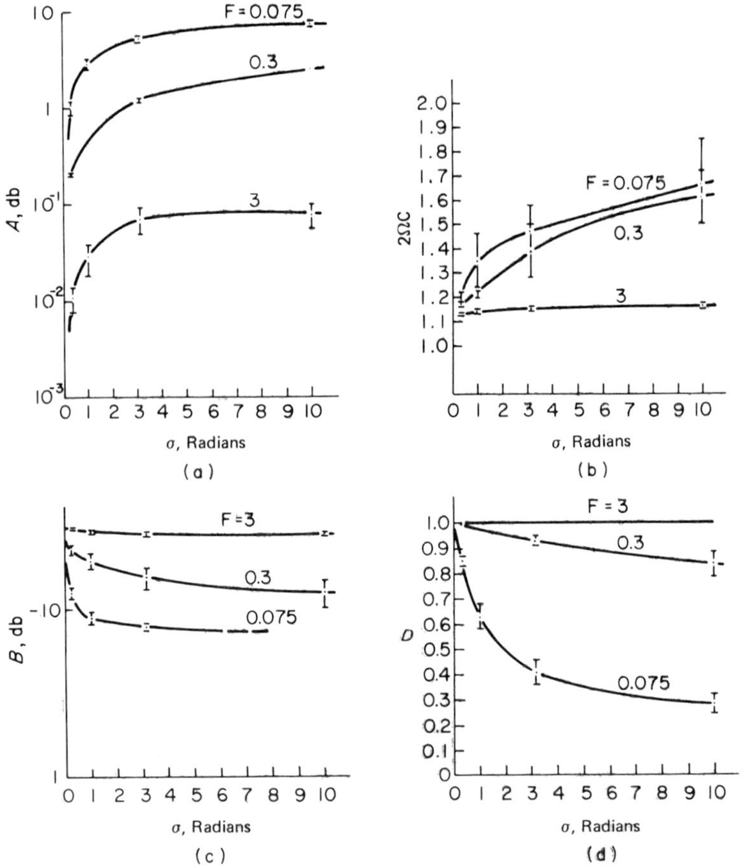

FIG. 6.4 The effect of Gaussian phase errors on resolution parameters. ($|B|^2$ was chosen to be a prolate spheroidal wave function, $c \approx 3$.) (a) The mean reduction in maximum A, as a function of the standard deviation, σ, of the phase error, with normalized correlation distance, F, as a parameter; (b) the mean resolution between 3 db points, as a function of the standard deviation, σ, of the phase error, with normalized correlation distance, F, as a parameter; (c) the mean sidelobe level, B, as a function of the standard deviation, σ, of the phase error, with normalized correlation distance, F, as a parameter; (d) the mean relative energy between first minima, D, as a function of the standard deviation, σ, of the phase error, with normalized correlation distance, F, as a parameter.

Analytic discussion. Again we see that, if the bandwidth of the phase error γ is large relative to $2\pi/X$, then the maximum excursions of γ must be kept small (with large probability) with respect to 1 rad. Therefore, an analytic investigation of random phase errors that assumes small variance broadband phase errors is of sufficient interest. We shall investigate the effect of these errors on resolution, as measured by the fraction of average energy in the desired resolution interval.

We have already noted the form of the energy $E_{I\gamma}$ in the resolution interval I, in the presence of phase errors γ in Eq. (6.4):

$$E_{I\gamma} \approx \int |B_y|^2 \, dy \int |B_\zeta|^2 \, d\zeta \, \frac{\sin \frac{1}{2}kI(y-\zeta)}{k(y-\zeta)} \exp[i(\gamma_y - \gamma_\zeta)].$$

The average, $E_{Im} \equiv E\{E_{I\gamma}\}$, then involves the characteristic function $\Phi_{\Delta\gamma}(1; y-\zeta)$ of the increment $\gamma(y) - \gamma(\zeta)$ evaluated at $+1$,

$$\Phi_{\Delta\gamma}(1; y-\zeta) \equiv E\{\exp i(\gamma_y - \gamma_\zeta)\}.$$

We assume $\Phi_{\Delta\gamma}$ to be a function only of the difference $y-\zeta$. Assuming small phase errors,

$$\exp i(\gamma_y - \gamma_\zeta) \approx 1 + i(\gamma_y - \gamma_\zeta) - \tfrac{1}{2}(\gamma_y - \gamma_\zeta)^2$$

and so

$$\Phi_{\Delta\gamma}(1; y-\zeta) \approx 1 - \sigma^2[1 - \rho(y-\zeta)], \tag{6.6}$$

where we have assumed that the mean, $E\{\gamma(y)\}$, of the phase error process $\{\gamma(x),\ x \in X\}$ is zero; ρ is, of course, the (normalized) covariance function of the phase error process.

We note the following facts which allow us a simple approximate calculation of E_{Im}:

1. $|B_y|^2$ is zero for $|y| > \frac{1}{2}X$ and is approximately a constant for $|y| < \frac{1}{2}X$;
2. $\frac{1}{2}kI = 2\pi I\lambda R_o$, where $I \approx \frac{1}{2}D_h$ so that $\frac{1}{2}kI \approx \pi/X$;

therefore, $|B_y|^2\,|B_\zeta|^2 \sin[\frac{1}{2}kI(y-\zeta)]/k(y-\zeta)$ is roughly a constant over the square $\mathscr{S}(X) \equiv X \times X$ in the (y, ζ)-plane. Thus, the value of the integral E_{Im}, and the value of the perturbation to E_{I0}, depends primarily on how $\rho(x)$ varies for $x \in X$. We now have the approximation

$$E_{Im} \approx E_{I0}\left\{1 - \frac{1}{X^2} \iint_{\mathscr{S}(X)} \sigma^2[1 - \rho(y-\zeta)] \, dy \, d\zeta\right\}. \tag{6.7}$$

In order to see the essential nature of the effect of these phase errors on the resolution measure E_{Im}, let

$$\rho(x) \equiv \begin{cases} 1, & |x| < \frac{1}{2}X_c, \\ 0, & |x| > \frac{1}{2}X_c. \end{cases}$$

Then it is easy to calculate that

$$E_{Im} \approx \begin{cases} E_{I0}, & X_c \geq \sqrt{2}X, \\ E_{I0}\left[1 - \sigma^2\left(1 - \dfrac{X_c}{\sqrt{2}X}\right)^2\right], & X_c < \sqrt{2}X, \quad \sigma < 1 \quad \text{rad}. \end{cases} \tag{6.8}$$

Equation (6.8) summarizes, in a simple manner, the essential dependence of resolution (energy measure) on the variance and relative bandwidth of small ($\sigma^2 \ll 1$) random phase errors. It is plotted in Fig. 6.5. We discuss the impact of this result on SAR system design in Sec. 6.6.

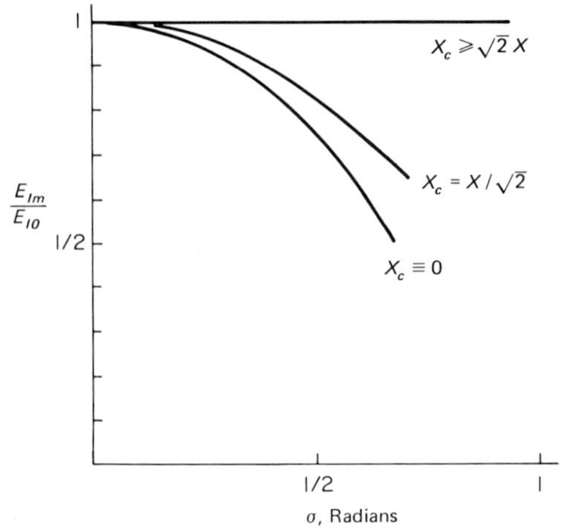

FIG. 6.5 The effect on resolution, measured by the fraction of energy of the system impulse response, of small ($\sigma < 1$ rad) random phase errors.

Average response. It is worth observing that the average response

$$E\{|q_\gamma(x)|^2\} = \int |B_y|^2 e^{-ikxy} \int |B_\zeta|^2 e^{ikx\zeta} \Phi_{\Delta\gamma}(1; y - \zeta)$$

from Eq. (6.2). For small phase errors, $\Phi_{\Delta\zeta}$ is given by Eq. (6.6), which we write in the form

$$(1 - \sigma^2) + R_\gamma(y - \zeta).$$

Thus, $E\{|q_\gamma(x)|^2\}$ can be expressed as two terms. The first is $(1 - \sigma)^2$ times the response when $\gamma \equiv 0$, $|q_0(x)|^2$. The second can be written as a convolution in the transform domain using Parseval's relation; since the Fourier transform of $|B(y)|^2 \exp(-ikxy)$ is just $q_o(\omega - kx)$ we have for the second term

$$\frac{1}{2\pi} |q_o|^2 * P_\gamma$$

where P_γ is the spectral density of $\{\gamma\}$—the Fourier transform of R_γ. So, for small random phase errors,

$$E\{|q_\gamma(x)|^2\} \approx (1 - \sigma^2) |q_0(x)|^2 + \frac{1}{2\pi} P_\gamma * |q_o|^2(x). \tag{6.9}$$

For reasons already mentioned, Eq. (6.9) is most meaningful when P_γ has bandwidth large relative to $2\pi/X$. Assuming this and noting that, in turn, $2\pi/X$ is itself much greater than the resolution interval (however reasonably defined) of $|q_0|^2$, we conclude that P_γ has a spread very large, compared to the approximate spread of $|q_0|^2$. Thus, $P_\gamma * |q_0|^2$ is a nearly constant background over the spread of $|q_0|^2$, generally contributing to a rise in sidelobe level. The first term indicates that the average peak response is decreased and hence also the signal-to-(additive) noise ratio.

6.3. EFFECTS ON LMVE SYSTEM

We have already noted that the MSE criterion is best used in conjunction with relatively large bandwidth phase errors, and we therefore make this assumption in this section. It is also indicated by now that the assumption of small phase errors is of sufficient interest, and we shall, in places, use this assumption also.

Referring to Fig. 6.6, we wish to calculate the $\mathrm{MSE}(h) \equiv E\{|g - \hat{g}|^2\}$ for

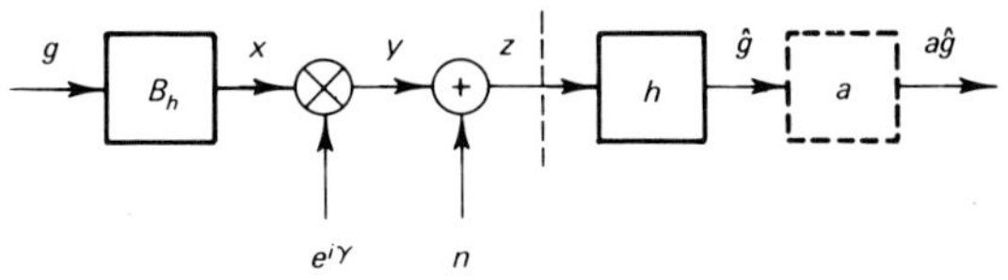

FIG. 6.6 Model for calculating MSE in a system with additive noise and phase errors.

an arbitrary processing filter h. $\{g\}$, $\{\gamma\}$, and $\{n\}$ are assumed to be independent, wide-sense stationary random processes with zero means and spectral densities P_g, P_s, and P_n, respectively. It is easy to calculate $\mathrm{MSE}(h)$ but first we observe that generally the (complex) amplitude scale of the image $\hat{g}$ is of no interest; therefore, we allow an operation that adjusts this amplitude scale independently of h (see Fig. 6.6). Now we are interested in the measure

$$\min_a \mathrm{MSE}\,(ah).$$

We quickly note that (completing the square)

$$\begin{aligned}\mathrm{MSE}(ah) &\equiv E\{|g - a\hat{g}|^2\} \\ &= \left| a(E\{|\hat{g}|^2\})^{1/2} - \frac{\overline{E\{g\bar{\hat{g}}\}}}{(E\{|\hat{g}|^2\})^{1/2}} \right|^2 + E\{|g|^2\} - \frac{|E\{g\bar{\hat{g}}\}|^2}{E\{|\hat{g}|^2\}}\end{aligned}$$

and hence that

$$a_{\text{opt}} = \frac{\overline{E\{g\bar{\hat{g}}\}}}{E\{|\hat{g}|^2\}}$$

and

$$\min_a \text{MSE}(ah) = E\{|g|^2\} - \frac{|E\{g\bar{\hat{g}}\}|^2}{E\{|\hat{g}|^2\}} .\dagger \tag{6.10}$$

We must now calculate the expectations appearing in Eq. (6.10); the details are relegated to Appendix 6A. The result is

$$\min_a \text{MSE}(ah) = \sigma_g{}^2 - |E\{e^{i\gamma}\}|^2 \cdot \frac{\left|\int \bar{\tilde{h}}\bar{\tilde{B}} P_g \, d\omega\right|^2}{\int |\tilde{h}|^2 \left[P_n + \dfrac{1}{2\pi} P_\varepsilon * (|\tilde{B}|^2 P_g)\right] d\omega} \tag{6.11}$$

where $\sigma_g{}^2$ is the variance of the process $\{g(x)\}$ and P_ε is the spectral density of the process $\{\exp i\gamma(x)\}$.

If we now assume small phase errors, $\exp i\gamma \approx 1 + i\gamma$ and

$$R_\varepsilon(x_1, x_2) \equiv E\{\exp i(\gamma_1 - \gamma_2)\} \approx 1 + R_\gamma(x_1 - x_2)$$

and, therefore

$$P_\varepsilon(\omega) \approx 2\pi\delta(\omega) + P_\gamma(\omega) \tag{6.12}$$

where δ is the "Dirac" (generalized) function. Thus, in Eq. 6.11) we can set

$$\frac{1}{2\pi} P_\varepsilon * (|\tilde{B}|^2 P_g) \approx |\tilde{B}|^2 P_g + \frac{1}{2\pi} P_\gamma * (|\tilde{B}|^2 P_g), \tag{6.13}$$

and also note that, to first order in σ^2, $|E\{e^{i\gamma}\}|^2 \approx 1 - \sigma^2$. Equation (6.11) then becomes

$$\min_a \text{MSE}(ah) \approx \sigma_g{}^2 - (1 - \sigma^2) \cdot \frac{\left|\int \bar{\tilde{h}}\bar{\tilde{B}} P_g \, d\omega\right|^2}{\int |\tilde{h}|^2 \left[P_n + |\tilde{B}|^2 P_g + \dfrac{1}{2\pi} P_\gamma * (|\tilde{B}|^2 P_g)\right] d\omega} \tag{6.14}$$

† The optimum filter h, or LMVE, is easily found from Eq. (6.10), using just the Schwarz inequality—see Chapter X.

Of special interest is the MSE of the system optimized for $\gamma \equiv 0$. The LIV filter for $\gamma \equiv 0$ was found in Sec. 4.3.2:

$$\tilde{h} = \frac{\bar{\tilde{B}} P_g}{P_n + |\tilde{B}|^2 P_g}. \tag{6.15}$$

We now note the following:

1. Generally speaking, $P_\gamma * (|\tilde{B}|^2 P_g)$ is of wider spread than $|\tilde{B}|^2 P_g$ and hence also, by Eq. (6.15), of wider spread than $|\tilde{h}|^2$. Therefore, it is not a bad approximation to replace $P_\gamma * (|\tilde{B}|^2 P_g)$ by its value—call it α—at $\omega = 0$ in Eq. (6.14).

2. The convolution $P_\gamma * (|\tilde{B}|^2 P_g)$, and therefore also α, are proportional to σ^2 so that if σ^2 is sufficiently small, we can write the denominator integral of Eq. (6.14) as two integrals $I_2 + I_3$, where $I_3 \ll I_2$, and I_3 is proportional to α.

3. If, therefore, we call the numerator integral of Eq. (6.14) I_1, then we have, to first order in σ^2,

$$\min_a \text{MSE}(ah) \approx \sigma_g^2 - \frac{|I_1|^2}{I_2} + \sigma^2 \frac{|I_1|^2}{I_2} + \frac{|I_1|^2 I_3}{I_2^2}. \tag{6.16}$$

4. Only I_3 depends on $\{\gamma\}$ and then only by the multiplicative factor α. We already have noted α is proportional to σ^2, but we now want to find out how it depends on the relative bandwidth of $\{\gamma\}$. To this end let

$$P_\gamma(\omega) \equiv 2\pi\sigma^2 \frac{\exp\left(-\frac{1}{2}\frac{\omega^2}{\Omega^2}\right)}{(2\pi)^{1/2}\Omega}$$

and

$$|\tilde{B}|\, P_g(\omega) \equiv 2\pi\nu^2 \frac{\exp\left(-\frac{1}{2}\frac{\omega^2}{\Delta^2}\right)}{(2\pi)^{1/2}\Delta}.$$

Then

$$\alpha \equiv P_\gamma * (|\tilde{B}|^2 P_g)|_{\omega=0} = \frac{(2\pi)\sigma^2\nu^2}{(\Omega^2 + \Delta^2)^{1/2}}.$$

The measure of bandwidth for $\{\gamma\}$ is, of course, Ω. Thus, setting $I_3 \equiv \alpha I_3'$ we have finally

$$\min_a \text{MSE}(ah) \approx \text{MSE}|_{\sigma^2 \equiv 0} + \sigma^2 \frac{|I_1|^2}{I_2} + \frac{(2\pi)^{1/2}\, \sigma^2\nu^2}{(\Omega^2 + \Delta^2)^{1/2}} \left(\frac{|I_1|^2 I_3'}{I_2}\right). \tag{6.17}$$

The result, Eq. (6.17) gives the general dependence of MSE on small large-bandwidth random phase errors $\{\gamma\}$ of variance σ^2 and spectral density bandwidth Ω. We observe:

1. The MSE increases linearly with σ^2 ($\sigma^2 \ll 1$).
2. The MSE decreases with increasing bandwidth Ω.

The second conclusion infers that, given a phase error process $\{\gamma\}$ with fixed variance σ^2, it affects the MSE less, the greater its bandwidth. This observation is compatible with our earlier assertion that the MSE will be large when very low frequency phase errors are present—linear phases, e.g. At the same time, this second conclusion is at odds with the resolution criterion, Eq. (6.8), for there, resolution *decreases* with increasing bandwidth.

6.4. EFFECTS ON AMBIGUITY FUNCTION

The discussion of Sec. 6.2 describes the effects of phase errors on adequately sampled signals, in particular the azimuth modulation of the SAR system. The effect of a linear phase error on sampled signals was discussed in Sec. 2.7 with particularization to the SAR azimuth modulation. It remains to describe the effect of relatively large bandwidth phase errors on the incompletely sampled azimuth signal, i.e., the SAR system azimuth ambiguity problem.

Instead of the azimuth modulation being representable as $B(x) \exp i\gamma(x)$ we now must assume the form

$$B(x)e^{i\gamma(x)} \sum h(x - nX_s).$$

After the usual matched (or phase-matched) filtering, we have the azimuth channel response as the summation (over n) of

$$q_{\gamma n}(x) = \frac{1}{2\pi} \int e^{iux} \tilde{B}(u - nu_s) \bar{\tilde{B}}_X(u)\, du, \qquad (n = 0, \pm 1, \ldots).$$

Proceeding just as before from Eq. (6.1) to Eq. (6.2), we find

$$q_{\gamma n}(x) \approx \int B_n(y) e^{i\gamma(y)} \exp[-i(kx - nu_s)y]\, dy$$

where

$$B_n(y) \equiv |B(y)| \cdot |B_X(y - nu_s)|,$$

assumed relatively independent of x for changes of x of the order of the resolution distance. Setting

$$\bar{x} \equiv x - nu_s/k = x - nX,$$

$$q_{\gamma n}(\bar{x}) = \int B_n(y) e^{i\gamma(y)} \cdot e^{-ik\bar{x}y}\, dy, \qquad n = 0, \pm 1, \pm \ldots,$$

describes the ambiguities about all the locations ($n = \pm 1, \pm 2, \ldots,$).

Suppose that we measure the ambiguity level by the energy of $q_{\gamma n}(\bar{x})$ for $x \in (-\frac{1}{2}I, \frac{1}{2}I)$: this criterion is less sensitive to small, relatively unimportant displacements of the location of the peak of the ambiguity. We then find the average energy in each ambiguous location, $E_{\mathrm{Im}}(n) \equiv E\{\mathscr{E}_{\mathrm{I}}(n)\}$,

$$\mathscr{E}_{\mathrm{I}}(n) \equiv \int_T |q_{\gamma n}(\bar{x})|^2 \, d\bar{x}, \qquad n = 0, \pm 1, \pm \ldots,$$

has for each n the form of that for $n = 0$—Eq. (6.4)—except that $|B|^2$ is replaced by $B_n(x) \equiv |B(x)| \cdot |B(x - nu_s)|$. But this implies that the behavior of $E_{\mathrm{Im}}(n)$ is much the same for any n: as the nature of the phase error process $\{\gamma(x), x \in X\}$ is changed, the $E_{\mathrm{Im}}(n)$ all change in the same general way. We may say that the azimuth ambiguity problem stays "internally" unchanged under broad bandwidth, small random phase errors.

6.5. EFFECT OF DISPERSIVE ERRORS

As we have already pointed out, the phase error analyses of Sec. 6.2 apply directly to the description of the effects of dispersive errors. For example, to describe the effects on the range channel of the SAR we replace $|B|^2$ everywhere by $|\tilde{F}|^2$, where F is the complex signal modulation, and set $k \equiv 1$; in particular the rough "thumb-rules" about tolerable errors carry over unchanged.

If linear FM is used, as it often is, then phase errors in the range-channel are analyzed exactly the same as in Secs. 6.2 and 6.3: $|B|^2$ is replaced by $|F|^2$, k by the linear FM rate of the range signal modulation, and X by the signal (single pulse) duration T.

6.6. BOUNDS ON PHASE ERRORS FOR SYSTEM DESIGN

We have seen that the effect of phase errors (and dispersive errors) on SAR systems designed for noise-free or additive noise conditions is very serious, generally speaking. As a rough rule, the phase error differences over the aperture, or signal duration (or spectrum for dispersive errors), should be kept small with respect to 1 rad.

As a specific example, in Chapter VIII, we will investigate at some length how to control the ambiguity (including sidelobe) level of the SAR system impulse response. We will see that c_h around 3 is a typical value of c_h when ambiguity level control around -30 db is being considered: from Fig. 8.3, we see that then we would be attempting to control the energy in the resolution cell to around 1%. From Eq. (6.8), we see that the phase errors must be controlled sufficiently well that

$$\sigma^2\left(1 - \frac{X_c}{(2X)^{1/2}}\right)^2 \leq 0.01. \tag{6.18}$$

In the worst case of very broad band errors, $X_c \approx 0$, we then must have

$$\sigma < 0.1 \text{ rad.} \tag{6.18a}$$

Qualitatively, this analytically derived result agrees with the numerical analysis result, Fig. 6.4d.

Part II: Motion Errors

6.7. INTRODUCTION

Among the remaining types of errors that we have not yet discussed, one of the most important for SAR systems is what we will call *motion errors*. This type of error arises when a scanning/sensing or recording beam does not move in uniform relative translational motion: then the desired convolution is not formed, but rather a perturbed convolution that is of the form

$$S_{r1}(y) = \int S_1(t)h[y - v_o t - \varepsilon_1(t)]\, dt. \tag{6.19}$$

In Eq. (6.19), $\varepsilon_1(t)$ is the perturbation from the uniform translational motion $v_o t$; e.g., $S_1(t)$ would be the signal to be recorded and S_{r1} would be the recorded signal.

Recalling the convolution that is performed by the SAR system as its physical beam senses the terrain, we see that any perturbations in the vehicle along-track velocity magnitude cause an error of this type. Barring events of cataclysmic consequence, these errors cannot arise in the other scanning direction, as the velocity of light is fixed. If recording operations are used, as they almost invariably are, then these errors do arise and are one of the primary design restraints for the recorders. For example, if a coherent optical processor is used, film velocities in both data and image recording cameras must be sufficiently uniform in their motion, and the CRT sweep velocity must also be sufficiently uniform.

Our purpose here is to analyze the effects of these motion errors in order to derive specifications on position and velocity errors in terms of signal bandwidth and time-bandwidth product. We shall generally assume the perturbation $\{\varepsilon_{1_t}\}$ is a random process (or variable) and shall investigate average behavior. We shall do this in the context of matched filtering, which is representative of the optimum processing procedures found in Chapter IV. (For concreteness, the reader may wish to visualize coherent optical processing, as described in Chapters I and V.)

Recording and Processing. As typical, consider the recording operation sketched in Fig. 6.7. Along the line $P(x)$, the recorder's spatial beam $h(x)$

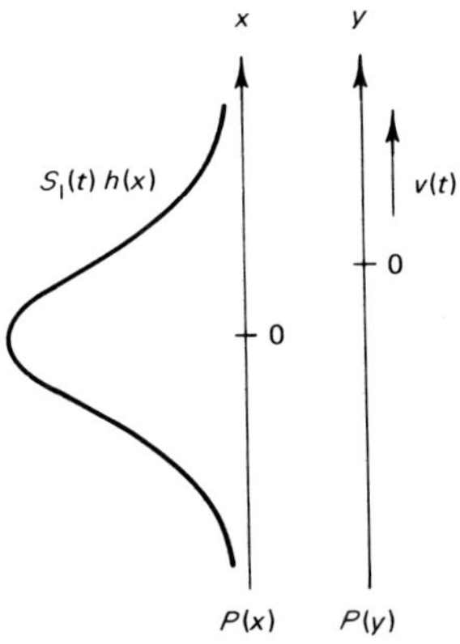

FIG. 6.7 The recording operation.

appears, modulated in amplitude by the temporal signal $S_1(t)$: the time-varying spatial distribution available for recording/sensing is then

$$S_2(t, x) \equiv S_1(t)h(x)$$

The recording medium moves along the line $P(y)$—essentially coincident with $P(x)$—in the x direction with velocity $v \equiv v_o + \varepsilon_v(t)$: thus

$$x = y - v_0(t - t_s) - \varepsilon_1(t)$$

where

$$\varepsilon_1(t) \equiv \int_{t_s}^{t} \varepsilon_v(t)\, dt.$$

By choosing the time scale, we may take $t_s \equiv 0$.

The incremental contribution to the record at any point $y \in P(y)$ is thus

$$dS_{r1}(y) = S_1(t)h(x)\, dt, \qquad x = y - v_o t - \varepsilon_1(t).$$

Integration gives Eq. (6.19). It is convenient to write this in spatial variables:

$$S_r(y) = \int_0^L S(z)h[y - z - \varepsilon(z)], \tag{6.20}$$

where

$$L \equiv v_o T,$$

$$S_r(y) \equiv v_o S_{r1}(y),$$

$$S(z) \equiv S_1\left(\frac{z}{v_o}\right),$$

and

$$\varepsilon(z) \equiv \varepsilon_1\left(\frac{z}{v_o}\right).$$

In order to analyze this random phenomenon, it is convenient to write Eq. (6.20) in the form

$$S_r(y) = \frac{1}{2\pi}\int d\omega\, e^{i\omega y}\tilde{h}(\omega)\int_0^L du\, S(u)\exp\{-i\omega[u + \varepsilon(u)]\}.$$

For now, when we take averages, the various characteristic functions of the process $\{\varepsilon\}$ arise naturally: they, of course, characterize the process completely.

Matched filter processing. As we said, we assume the processor to be an LIV filter, $k(y)$, matched to the recorded signal when $\varepsilon \equiv 0$ and $h(x) = \delta(x)$—the unit impulse generalized function:

$$k(y) = \bar{S}(-y).\dagger$$

The output of the processor is then $|g(y)|^2$ where

$$\begin{aligned} g(y) &\equiv S(y) * k(y) \\ &= \frac{1}{2\pi}\int d\omega\, e^{i\omega y}\overline{\tilde{S}(\omega)}\tilde{h}(\omega)\int_0^L du\, S(u)\exp\{-i\omega[u + \varepsilon(u)]\}. \end{aligned} \tag{6.21}$$

The output when $\varepsilon \equiv 0$ and $h(x) = \delta(x)$, we shall call the "ideal output," $|g_I|^2$, where

$$g_I(y) \equiv \frac{1}{2\pi}\int d\omega\, e^{i\omega y}|\tilde{S}(\omega)|^2. \tag{6.22}$$

The output when $\varepsilon \equiv 0$, we shall call the "unperturbed output," $|g_o|^2$, where

$$g_o(y) \equiv \frac{1}{2\pi}\int d\omega\, e^{i\omega y}|\tilde{S}(\omega)|^2\,\tilde{h}(\omega). \tag{6.23}$$

† As always for the SAR application, we need linear, quasi-invariant filters; recall the discussion of Sec. 4.1.

This has the simple interpretation as the output of a cascade of the matched filter and the recorder impulse response with the spatial signal S as input. These situations are shown in Fig. 6.8.

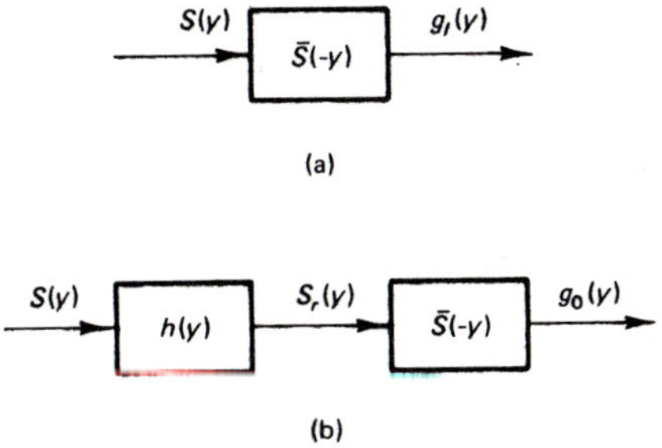

FIG. 6.8 The processor output when: (a) there are no motion errors, and $h(x) \approx \delta(x)$; (b) when there are no motion errors.

The average output. The expected output is

$$E\{|g(y)|^2\} = \frac{1}{4\pi^2} \int d\omega_1 \, e^{i\omega_1 y} \bar{\tilde{S}}(\omega_1)\tilde{h}(\omega_1) \int_0^L du_1 S(u_1) e^{-i\omega_1 u_1}$$
$$\cdot \int du_2 \, e^{-i\omega_2 y} \tilde{S}(\omega_2)\bar{\tilde{h}}(\omega_2) \int_0^L du_2 \, \bar{S}(u_2) e^{i\omega_2 u_2}$$
$$\cdot \, E\{\exp[-i\omega_1 \varepsilon(u_1) + i\omega_2 \, \varepsilon(u_2)]\}. \tag{6.24}$$

The expectation, E, appearing on the right side of this last equation, we recognize as the joint characteristic function

$$\phi_{\varepsilon_1, \varepsilon_2}(-\omega_1, \omega_2 \, ; u_1, u_2)$$

of the random variables $\varepsilon(u_1)$, $\varepsilon(u_2)$. We have encountered this form in the discussions of phase errors; there are superficial similarities between these two types of errors.

6.8. LARGE BANDWIDTH MOTION ERRORS

When the process $\{\varepsilon_x\}$ fluctuates many times over the distance L, then it is reasonable to assume that over most of the (u_1, u_2)-plane, $\varepsilon(u_1)$ and $\varepsilon(u_2)$ are independent random variables. Then in Eq. (6.24) we may make the approximation

$$\varphi_{\varepsilon_1, \varepsilon_2}(-\omega_1, \omega_2 \, ; u_1, u_2) \approx \varphi_{\varepsilon_1}(-\omega_1; u_1) \cdot \varphi_{\varepsilon_2}(\omega_2 \, ; u_2),$$

where, e.g., $\varphi_{\varepsilon 1}$ is the characteristic function of the random variable $\varepsilon(u_1)$. Now Eq. (6.24) becomes

$$E\{|g(y)|^2\} \approx |E\{g(y)\}|^2$$

where

$$E\{g(y)\} = \frac{1}{2\pi} \int d\omega \, e^{i\omega y} \bar{\tilde{S}}(\omega) \tilde{h}(\omega) \int_0^L du \, e^{-i\omega u} S(u) \bar{\varphi}_{\varepsilon_u}(\omega; u).$$

If the error process has a stationary first order distribution, then the characteristic function does not (by definition) depend on u, and we have the very simple result that the mean output is $|g_m|^2$, where

$$g_m(y) \equiv E\{g(y)\} = \frac{1}{2\pi} \int d\omega \, e^{i\omega y} \bar{\tilde{S}}(\omega) \tilde{h}(\omega) \tilde{S}(\omega) \bar{\varphi}_\varepsilon(\omega). \tag{6.25}$$

This result can be interpreted as the output of a cascade of three LIV filters—the recorder impulse response, the matched filter, and a filter whose impulse response is essentially the univariate density function, $f_\varepsilon(-y)$, of the error process $\{\varepsilon_y\}$—with the spatial signal S as input.† See Fig. 6.9 and compare Fig. 6.8.

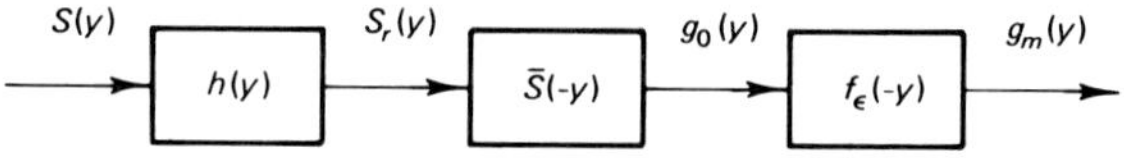

FIG. 6.9 The average processor output for large bandwidth, stationary motion errors.

The result, Eq. (6.25), has other interesting implications. For example, we can measure the average frequency response of a system with large bandwidth motion errors using sinusoids in a fairly conventional way—we need only find the average response for each frequency. Having this average frequency response, we can then calculate the expected effect of the motion errors on any specified signal.

As another example, suppose that we wish to maintain a certain average system resolution, as measured by radius of gyration. Knowing $|g_I|^2$ or $|g_o|^2$ and how much degradation in resolution we are willing to allow in $|g_m|^2$, we can place a restraint on the first-order distribution of this type of motion error. For example, suppose g_I is a Gaussian-shaped pulse of radius of gyration (resolution) v, and that $\{\varepsilon_y\}$ has a Gaussian first-order distribution of variance σ_ε^2: then g_m is a Gaussian-shaped pulse of resolution $(v^2 + \sigma_\varepsilon^2)^{1/2}$. If, e.g., negligible increase in resolution is to occur, we must have

$$\frac{\sigma_\varepsilon^2}{v^2} \ll 1. \tag{6.26}$$

† By definition, the characteristic function and the probability density function (assuming it exists) are a Fourier transform pair.

That is, the standard deviation of the position error must be a small part of the system resolution—a not unexpected result.

Velocity errors. If the source of position errors is attributable to velocity errors $\{\varepsilon_{v_t}\}$, then the position error

$$\varepsilon(y) = \int_0^{y/v_o} \varepsilon_v(t)\, dt.$$

This is a nonstationary process, and generally ε_{u_1} and ε_{u_2} are correlated:

$$R_\varepsilon(y_1, y_2) = \int_0^{y_1} du \int_0^{y_2} dv\, R_v(u - v).$$

It seems intuitively clear that we will get a pessimistically large effect if we assume $\{\varepsilon_v\}$ to be of broad bandwidth, with variance equal to the maximum

$$\sigma_\varepsilon^2(L) = R_\varepsilon(L, L) = \int_0^T du \int_0^T dv\, R_{\varepsilon_v}(u - v).$$

If $\{\varepsilon_v\}$ has variance σ_v^2 and correlation distance $T_v \ll T$, then

$$\sigma_\varepsilon^2(L) \approx \sigma_v^2 T T_v.$$

Substituting this in Eq. (6.26) gives the condition

$$\frac{\sigma_v^2 T T_v}{v^2} \ll 1. \tag{6.27}$$

We can put Eq. (6.27) into an alternate form. Since v is a measure of system spatial resolution, $\Omega_y \equiv v^{-1}$ is a measure of system spatial bandwidth, and $\Omega \equiv v_o \Omega_y$ is a measure of the temporal bandwidth of the signal S. Also, $\Omega_v \equiv 1/T$ is a measure of the bandwidth of the spectral density of the velocity error process $\{\varepsilon_{vt}\}$. Substituting, Eq. (6.27) becomes

$$\frac{\sigma_v^2}{v_o^2} \ll \frac{\Omega_v}{\Omega} \cdot \frac{1}{T\Omega}. \tag{6.27a}$$

Thus, e.g., the allowable percent velocity error—the left side of Eq. (6.27a)—is inversely proportional to the time-bandwidth product of the signal to be processed.

6.9. SMALL BANDWIDTH MOTION ERRORS

An especially prevalent type of motion error in mechanical recording mechanisms—e.g., camera drives—is due to eccentric bearings, gears, etc. A simple model appropriate to such errors is

$$\varepsilon(y) \equiv \varepsilon \cos(\omega_o y + \theta),$$

where ε and θ are random variables, the latter uniformly distributed over $[0, 2\pi]$.

It is convenient to employ Eq. (6.21) prior to substitution; we use the Jacoby-Anger formula (see Chapter VII):

$$\exp[i\omega\varepsilon \cos(\omega_o u + \theta)] = \sum_{m=0}^{\infty} \delta_m i^m J_m(\omega\varepsilon)\cos m(\omega_o u + \theta).$$

Then

$$g(y) = \frac{1}{4\pi} \sum_{m=0}^{\infty} \delta_m i^m \int d\omega\, e^{i\omega y}\overline{\tilde{S}(\omega)}\tilde{h}(\omega)J_m(\varepsilon\omega)$$
$$\cdot [\tilde{S}(\omega + m\omega_o)e^{im\theta} + \tilde{S}(\omega - m\omega_o)e^{-im\theta}]. \tag{6.28}$$

Although Eq. (6.28) is of rather forbidding appearance, if we keep our wits about us, the general effects are easily seen; for in Chapter II, especially Sec. 2.7, we saw what the effects of displaced signal spectra are: loss of resolution, background (dispersed) clutter, and false images. Equation (6.28) tells us that if it is possible, the mechanical apparatus (e.g.) should be chosen so that all harmonic frequencies ω_o fall above the spectrum of the signal.

High frequency perturbations. If we do manage to have the harmonics ω_o all sufficiently high, then only the term $m = 0$ in Eq. (6.28) is nonzero:

$$E\{|g(y)|^2\} = \frac{1}{4\pi^2} \int d\omega_1 e^{i\omega_1 y} |\tilde{S}(\omega_1)|^2 \bar{\tilde{h}}(\omega_1)$$
$$\cdot \int d\omega_2\, e^{-i\omega_2 y} |\tilde{S}(\omega_2)|^2 \tilde{h}(\omega_2)$$
$$\cdot\, E\{J_o(\varepsilon\omega_1)J_o(\varepsilon\omega_2)\}.$$

Recalling that $J_o(0) = 1$, and $J_o(1) \approx 0.77$, if ε is much less than the reciprocal of the signal spatial bandwidth (with large probability), then little distortion of the output occurs:

$$\frac{{\sigma_\varepsilon}^2}{\nu^2} \ll 1.$$

For velocity errors, we get the condition

$$\frac{\sigma_{\varepsilon_v}^2}{{v_o}^2} \ll \frac{{\omega_o}^2}{\Omega^2}, \qquad \omega_o > 2\pi\Omega.$$

To see the first order effect of such an error, we can expand the Bessel functions and retain terms to order two in ε:

$$E\{|g(y)|^2\} \approx |g_o(y)|^2 + {\sigma_\varepsilon}^2 \operatorname{Re}\left\{\bar{g}_o(u) \frac{d^2}{dy^2} g_o(y)\right\}. \tag{6.29}$$

The general effect is seen to be the increase in level and fluctuation rate of the sidelobe structure. As σ_ε^2 increases, the main lobe of the response $E\{|g(y)|^2\}$ begins to be affected, at first "tearing," and then dispersing. (These effects can be seen by sketching simple examples.)

Intraband frequency perturbations. If the natural spatial frequency ω_o of the sinusoidal motion error falls within the signal spectral band, greater distortion can occur. To find the first-order effects on the average output, we can again expand the Bessel functions. The form of the expectation is greatly simplified because of the assumed distribution on θ. The result is the right side of Eq. (6.29), plus the term

$$\frac{\sigma_\varepsilon^2}{4}\left\{\left|\frac{d}{dy}(\bar{s}(-y) * h(y) * [s(y)e^{-i\omega_o y}])\right|^2 + \left|\frac{d}{dy}(\bar{s}(-y) * h(y) * [s(y)e^{i\omega_o y}])\right|^2\right\}. \tag{6.30}$$

The effects of this term are like those described for high frequency perturbations, except additionally broad clutter and false peaks can occur. This is most easily seen, if we note that $s(y)\exp \pm i\omega_o y$, the input signal or data, appears to the processing structure, $\bar{s}(-y) * h(y)$, on an incorrect center frequency: thus, we can expect the effects described in Sec. 2.7.

Very low frequency perturbations. As an example of very low frequency motion errors, suppose that there is an unknown constant velocity error ε_v. Such an error is correctable: we wish to see how well we must know the velocity. From Eq. (6.21) with $h(y) \equiv \delta(y)$,

$$g(y) = \int_0^L dz\, \bar{s}(-z)s\left[(1-\varepsilon')\left(\frac{y}{1-\varepsilon'} - z\right)\right],$$

where $\varepsilon' \equiv \varepsilon_v/v_o$. In addition to the expected scale change, we note that $s\{(1-\varepsilon')z\}$ appears instead of $s(z)$. Let s be a linear FM wave with FM rate $2\pi\Omega/Tv_o^2$ and—in view of the results of Sec. 6.2.1—restrict the quadratic phase error to less than $\pi/2$ rad. Then we must have

$$\frac{\varepsilon' k L^2}{4} < \frac{\pi}{2},$$

or

$$\varepsilon' \equiv \frac{\varepsilon_v}{v_o} < \frac{1}{\Omega T}.$$

Thus, the relative constant velocity error must be less than the reciprocal of the time-bandwidth product of the signal to be processed.

APPENDIX 6A: CALCULATION OF SOME EXPECTATIONS

Referring to Sec. 6.3, especially Eq. (6.10) and Fig. 6.6, we want to calculate the averages $E\{|g|^2\}$, $E\{g\bar{\hat{g}}\}$, and $E\{|\hat{g}|^2\}$.

Of course, $E\{|g|^2\}$ is just the variance of the process $\{g(x)\}$ and, if P_g is the process spectral density, then by the latter's definition

$$E\{|g|^2\} = \frac{1}{2\pi}\int P_g(\omega)\, d\omega. \tag{6A.1}$$

To calculate $E\{|\hat{g}|^2\}$, we have first that, if P_z and P_g are the spectral densities of random processes $\{z(x)\}$ and $\{\hat{g}(x)\}$, and $\hat{g} = h * z$, then

$$P_{\hat{g}}(\omega) = |\tilde{h}(\omega)|^2\, P_z(\omega). \tag{6A.2}$$

Referring to Fig. 6.6 we have $P_z = P_y + P_n$ by the independence of $\{n\}$, $\{\gamma\}$, and $\{g\}$, which also gives $P_y = (1/2\pi)\, P_\varepsilon * P_z$ where $\varepsilon \equiv \exp i\gamma$. But the fact that gave us Eq. (6A.2), also gives us $P_x = |\tilde{B}|^2\ P_g$. Combining all these facts, we have

$$\begin{aligned} E\{|\hat{g}|^2\} &\equiv \frac{1}{2\pi}\int P_{\hat{g}}(\omega)\, d\omega \\ &= \frac{1}{2\pi}\int |\tilde{h}(\omega)|^2 \left\{ P_n(\omega) + \frac{1}{2\pi} P_\varepsilon(\omega) * [|\tilde{B}(\omega)|^2\, P_g(\omega)] \right\} d\omega. \end{aligned} \tag{6A.3}$$

Finally, we have

$$\begin{aligned} E\{g\bar{\hat{g}}\} &= E\{g \cdot \overline{h * [n + e^{i\gamma}(B * g)]}\} \\ &= E\{e^{-i\gamma}\} E\{g(\overline{h * B * g})\}. \end{aligned}$$

The second expectation is easily found by writing out the convolution and using Parseval's relation: thus,

$$E\{g\bar{\hat{g}}\} = E\{e^{-i\gamma}\} \cdot \frac{1}{2\pi}\int \bar{\tilde{h}}\bar{\tilde{B}} P_g\, d\omega. \tag{6A.4}$$

Substituting Eqs. (6A.2), (6A.3), and (6A.4) into Eq. (6.10) gives Eq. (6.11).

COMMENTS AND REFERENCES

Phase errors are as ubiquitous as death, taxes, and additive noise; not surprisingly, so is the analysis of them, though to a lesser extent. We do not propose to offer a bibliography, any more than we would for the analysis of effects of additive noise. The effects of phase error, especially quadratic errors, were discussed, in the specific context of linear FM modulations, by Klauder *et al.* (1960). Numerical analyses of effects of phase errors on the Fourier transform have been done by Greene and Moller (1962), and Harger and Crimmins (1965). Relevant analyses on the same problems with resolution as a criterion, especially radius of gyration, was discussed in a series of papers by Brown and Palermo; see, e.g., Brown and Palermo (1963a), Brown and Palermo (1965), and Brown (1967). The analysis of the effects on phase errors with MSE criterion has been discussed by Brown and Palermo (1962), and Harger (1967a). The reolution measure of energy in this context is novel.

The discussion of Part II follows Harger (1965a).

VII

Nonlinearities

7.1. INTRODUCTION

The signal amplitudes encountered in synthetic aperture radar systems have a very large range of possible values, generally much greater than the linear range of practical radar receiving and processing components. It is, therefore, of considerable interest to have some knowledge of the effects of nonlinearities on system performance.

The "no-memory" nonlinearity is one of the few types of nonlinearities sufficiently well understood to enable quantification of its effects: a "no-memory" nonlinearity has output $f(x_t)$ for input x_t, where f is an ordinary real function of a real variable. Fortunately, this type of nonlinearity model satisfactorily explains the major effects that are observed in these systems. Most RF and IF limiters have large bandwidth relative to the signal bandwidth and so are approximately "memoryless." However, some types, such as traveling wave tube limiters, may not fit this model very well; their effects are not understood sufficiently well at present to predict their effects in these systems. One of the most severely limited ranges occurs in the recording operation, e.g., the CRT-film system. These systems can be adequately modeled by a no-memory nonlinearity with isolated pre- and post-filters.

Furthermore, most analyses assume a narrow-band signal input to the nonlinearity, so that subsequent filtering can eliminate much of the harmonics and intermodulations present in the output. Except for Sec. 7.4, we shall always make this assumption. In practice, this is a good assumption for limiting prior to recording; since recording bandwidth comes at a premium, the signals at that point in the system are not narrow-band.

7.2. THE TRANSFORM METHOD

The nonlinear characteristics generally encountered in these systems are modelled by a symmetrical "smooth limiter," as shown in Fig. 7.1. In the case of film, this is only approximately true, as it responds only to magnitude, the signal therefore being normally recorded on a bias. A convenient function to use for the nonlinearity is essentially the "error function"

$$f(x) = \frac{L}{\theta}\left(\frac{2}{\pi}\right)^{1/2} \int_0^x \exp\frac{-\lambda^2}{2\theta^2}\, d\lambda. \tag{7.1}$$

L, of course, is the limiting level, and θ varies, roughly speaking, the range of the linear region. It can be shown that the properties we shall be

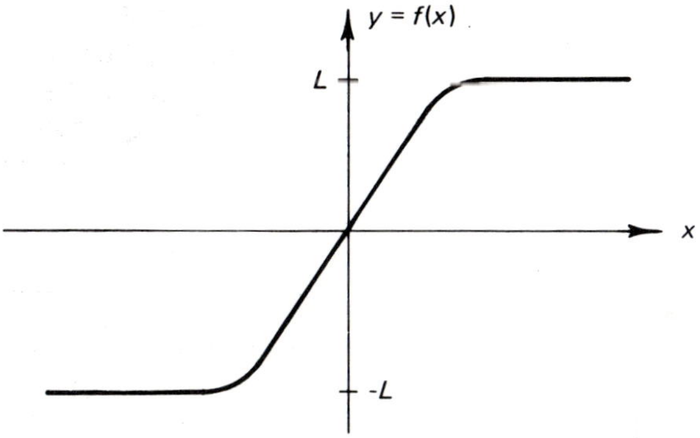

FIG. 7.1 General nonlinearity encountered in synthetic aperture radar systems.

interested in here are fairly insensitive to the exact form of the limiter. The limiting cases are, for θ very large, linear operation, and for θ very small, "ideal" limiting:

$$\lim_{\theta\to 0} f(x) = L \operatorname{sgn} x \equiv \begin{cases} +L, & x > 0, \\ 0, & x = 0, \\ -L, & x < 0. \end{cases} \tag{7.2}$$

The most successful method of analyzing no-memory nonlinearities is known as the "transform" or "characteristic function" method and is based on the observation that we can write

$$y_t = f(x_t) = \frac{1}{2\pi}\int e^{iux_t}\tilde{f}(u)\, du, \tag{7.3}$$

where, e.g., $\tilde{f}$ is the Fourier transform of f: the random phenomenon is now more explicitly displayed, and any subsequent averaging operations lead to various characteristic functions of the $\{x_t\}$ process. The functions, f, we wish to consider here—such as the error function—do not, however, possess ordinary Fourier transforms. The best way around this problem appears to be through the use of generalized functions and generalized Fourier transforms;

the characteristic function method has been established for present purposes for such transforms.

The generalized Fourier transform of the error function can be formally derived easily by noting that it is the integral of the form $a \exp -bx^2$ which has the well-known (ordinary) Fourier transform $\propto \exp -\beta u^2$; therefore, the (generalized) Fourier transform of the error function is formally $\propto [\exp - \beta u^2]/iu$ by the rules of ordinary Fourier transforms. In fact, we find

$$\tilde{f}(u) = 2L \frac{\exp[-\frac{1}{2}\theta^2 u^2]}{iu}. \tag{7.4}$$

For the "ideal" limiter we have

$$\lim_{\theta \to 0} \tilde{f}(u) = \frac{2L}{iu}. \tag{7.5}$$

Since the signal input to the nonlinear device is narrow-band, it is very convenient to employ a bandpass representation—the "envelope and phase" form—for the signal:

$$x(t) = V_t \cos(\omega_0 t + \psi_t), \tag{7.6}$$

where V_t is the envelope, ψ_t the phase, and ω_o the carrier frequency of the wave. We have now

$$y_t = \frac{1}{2\pi} \int \exp[iuV_t \cos(\omega_0 t + \psi_t)] \tilde{f}(u)\, du.$$

We now use an expansion that is extremely useful in the transform method, the Jacoby-Anger formula

$$\exp[i\alpha \cos \beta] = \sum_{m=0}^{\infty} \varepsilon_m i^m J_m(\alpha) \cos m\beta, \tag{7.7}$$

where $\varepsilon_m = 1$, if $m = 0$, and $\varepsilon = 2$, otherwise, and J_m is the usual mth order Bessel function of the first kind. Thus

$$\begin{aligned} y_t &= \frac{1}{2\pi} \int du\, \tilde{f}(u) \exp[iuV_t \cos(\omega_o t + \psi_t)] \\ &= \sum_{m=0}^{\infty} \varepsilon_m i^m \int J_m(uV_t) \tilde{f}(u)\, du \cdot \cos[m(\omega_o t + \psi_t)]. \end{aligned} \tag{7.8}$$

If we assume that x_t has a very narrow bandwidth, each of the integrals is also of narrow bandwidth: we have, therefore, remarkably simply obtained a spectral decomposition of the output of a no-memory nonlinearity. If we follow the nonlinearity with a narrow-band filter centered on ω_o—the location of the desired output—then the output of the filter is approximately

$$z(t) = \frac{i}{\pi} \int J_1(uV_t) \tilde{f}(u)\, du \cdot \cos(\omega_o t + \psi_t). \tag{7.9}$$

Examining Eq. (7.9) we see that we have established an important result:

> The phase of a narrow-band waveform is approximately unchanged by passage through a no-memory nonlinearity followed by a narrow-band filter.

This fact is, of course, very important in practical synthetic aperture systems. If the phase were not preserved, the effects could be disastrous upon the radar's operation: (1) the phase of the azimuth histories would be altered, degrading or even preventing compression; (2) the phase of the signal would be altered, degrading or even preventing compression, if dispersed pulses were used.

It is not to be concluded from this result that there is no distortion beyond loss of envelope. For, in the multiple input situation, the phase ψ_t depends in a complicated way upon the phases and amplitudes of the individual components. Also, there is always noise present. Thus, we shall have to examine such cases more carefully.

Using Eq. (7.4) for the transform of the limiter characteristic, Eq. (7.9) becomes

$$z(t) = \frac{2L}{\pi} \int u^{-1} J_1(V_t u) \exp(-\tfrac{1}{2}\theta^2 u^2)\, du \cdot \cos(\omega_o t + \psi_t).$$

The integral is evaluated by the formula of Weber and Sonine to give

$$z(t) = \begin{cases} \left(\dfrac{2}{\pi}\right)^{1/2} \dfrac{L}{\theta}\, {}_1F_1\left(\tfrac{1}{2}; 2; -\dfrac{V_t^2}{2\theta^2}\right) \cdot \cos(\omega_o t + \psi_t), & V_t > 0, \\ 0, \quad V_t = 0, & \end{cases} \tag{7.10}$$

where ${}_1F_1$ is the confluent hypergeometric function. It is of interest to examine the asymptotic limits

$$z(t) \approx \begin{cases} \left(\dfrac{2}{\pi}\right)^{1/2} \dfrac{L}{\theta} \cdot V_t \cos(\omega_o t + \psi_t), & \dfrac{V_t^2}{2\theta^2} \ll 1, \quad \text{(7.11a)} \\ \dfrac{4L}{\pi} \cdot \cos(\omega_o t + \psi_t), & \dfrac{V_t^2}{2\theta^2} \gg 1. \quad \text{(7.11b)} \end{cases}$$

Equation (7.11a) describes operation in the linear region and Eq. (7.11b) describes operation for "ideal," or hard, limiting.

We shall continue with the discussion for the hard limiting case, as described by Eq. (7.11b), as most analysis has been confined to this case. To obtain results in other than a few limiting cases, rather arduous or machine computation is required.

A problem encountered in analyzing the effects of nonlinearities is the identification of the signal in the output. Usually implicit in the identification is the type, if any, of processing that the output subsequently undergoes. The characteristic function method identifies certain averages—specifically, covariance functions—as signals. But this can be meaningless in systems that do not perform averaging; this will be discussed further in Sec. 7.3.2. Having identified the output signal, the signal-to-noise ratio is generally computed to characterize the effects of the nonlinearity, as we shall do in Sec. 7.3. Again, the signal-to-noise ratio may or may not be meaningful, depending on the subsequent processing. In Sec. 7.4, we shall adopt a mean-square error criterion.

7.3. MULTIPLE SIGNALS AND NOISE: HARD LIMITING

We shall further examine these results, by assuming the input to be two terms, $x_t = s_t + n_t$, where

$$s_t \equiv P_t \cos(\omega_o t + \theta_t),$$

$$n_t = N_t \cos(\omega_o t + \varphi_t).$$

As yet, we will not specify further these two components, except that we will have to assume one, say s_t, is generally much larger than the other, in order to get simple results. We recall we are also proceeding under the assumption of hard limiting.

Now we have

$$V_t = [(P_t \cos \theta_t + N_t \cos \varphi_t)^2 + (P_t \sin \theta_t + N_t \sin \varphi_t)^2]^{1/2}$$

and

$$\psi_t = \arctan\left(\frac{P_t \sin \theta_t + N_t \sin \varphi_t}{P_t \cos \theta_t + N_t \cos \varphi_t}\right).$$

With the assumption that nearly always $P_t \gg N_t$, we have

$$\psi_t \approx \theta_t + \frac{N_t \sin(\varphi_t - \theta_t)}{P_t}, \tag{7.12}$$

as is easily seen from Fig. 7.2. Then, by Eqs. (7.11b) and (7.12) we have, upon expansion

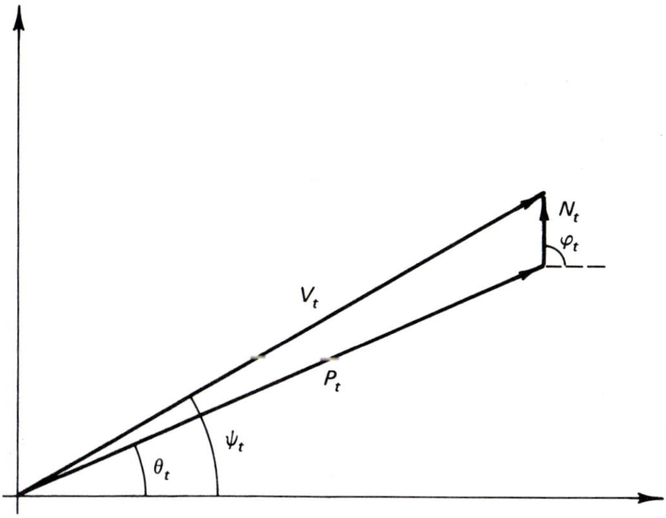

FIG. 7.2 Phase relations for resultant phase angle of two wave-forms, one relatively strong.

$$z(t) \approx \frac{4L}{\pi} \cos(\omega_o t + \theta_t) - \frac{4L}{\pi} \frac{N_t}{2P_t} \cos(\omega_o t + \varphi_t) + \frac{4L}{\pi} \frac{N_t}{2P_t} \cos(\omega_o t + 2\theta_t - \varphi_t). \tag{7.13}$$

We can tentatively identify each of these three terms:

1. The first term is an output due to the input s_t and is identified with s_t,† though the amplitude modulation has been lost.
2. The second term is an output due to the input term n_t and is identified with n_t, though the amplitude modulation may be modified.
3. The third term is an output due to both inputs and is termed an intermodulation product.

We note the following important result:

The ratio of the envelope of the output component identified with the strong input term, to the envelope of the output component identified

† That is to say, if s_t were the input signal, then the first term would be called the "output signal."

with the weak input term, is 6 db greater than the ratio of the envelope of the strong input term, to the envelope of the weak input term.

This is the famous *weak signal suppression* effect.

We shall take the result, Eq. (7.13) and apply it to situations that arise in the synthetic aperture radar application. These are:

1. a strong signal and a weaker ensemble of signals and/or noise;
2. a weak signal and a strong ensemble of signals and/or noise;
3. two strong signals, one relatively weaker.

7.3.1. Strong Signal, Weak Background

We consider first the case when s_t represents a strong signal waveform, and n_t represents a weaker waveform. There are two interesting ways to take n_t: (1) it can represent random noise (receiver, galactic, etc.); (2) it can represent an ensemble consisting of great number of weak signals—e.g., all the returns from other scatterers. This situation arises relatively infrequently with synthetic aperture signals because, as one would expect, it is a low probability event that a single scatterer's returned signal can be much larger than the sum of the thousands of other scatterer's returned signals. Nevertheless, it is an important phenomenon for these systems.

In these systems, the signal envelope, P_t, is approximately either a constant or zero; this will be kept in mind as we proceed. From Eq. (7.13) we see that the output signal (the first term) is stripped of its envelope, or amplitude modulation. Thus, in the output image of the radar, strong signals will tend to have the familiar sin t/t envelope despite any antenna pattern and signal spectral weighting that might have been designed. Conversely, if n_t represents the ensemble of weak targets, then their output, the second term of Eq. (7.13), approximately preserves its amplitude modulation.

As already observed, the ensemble of signals is suppressed —6 db relative to the strong target. This effect is occasionally observed dramatically in synthetic aperture radar images. For example, suppose the radar vehicle has a velocity parallel to a railroad track, which is a very strong scatterer, and the radar employs a dispersed signal modulation. Then beyond the railroad track for a distance equal to the dispersed pulse length, an apparently very good image will appear, but it will be reduced in intensity level.

The fact that the suppressed image is apparently very good implies that the intermodulation, the third term of Eq. (7.13), has negligible effect on the image. The reason for this is that, though amplitude of the intermodulation term is equal to the amplitude of the weak ensemble term, its phase is such that, generally, it will not compress upon passage through the processor.

The discussion of the last paragraph implies still another important fact:

an ensemble of undesired waveforms (intermodulation products, e.g.), even though of energy equal to an ensemble of desired waveforms, can have negligible effect upon output image quality. These remarks have a direct bearing upon the ambiguity problem, as we will discuss further in Chapter IX.

Weak noise background. If n_t is random noise, then it is customary to call output noise both the second and third terms of Eq. (7.13). Intuitively, we can see that, since the phases of these two terms are different, their amplitudes rarely will add—this property between two waveforms is called "incoherency"—but rather their intensities, or power, will add. The resultant power of the output noise is therefore -3 db relative to the output signal.

Perhaps the best way to establish this result is as follows. Writing Eq. (7.13) in the form

$$z(t) \approx \frac{4L}{\pi} \cos(\omega_o t + \theta_t) + \frac{4L}{\pi} \frac{N_t \sin(\varphi_t - \theta_t)}{P_t} \sin(\omega_o t + \theta_t),$$

the power of the signal term is one half the power of its envelope, i.e., $\frac{1}{2}(4L/\pi)^2$. The mean, or average, power of the noise term is one half the power of its envelope:

$$\left(\frac{4L}{\pi}\right)^2 \frac{1}{2P_t^2} E\{[N_t \sin(\varphi_t - \theta_t)]^2\} = \left(\frac{4L}{\pi}\right)^2 \frac{\sigma^2}{2P_t^2},$$

where we have used the facts†

$$E\{(N_t \sin \varphi_t)^2\} = E\{(N_t \cos \varphi_t)^2\} = E\{n_t^2\} \equiv \sigma^2,$$

$$E\{N_t \sin \varphi_t \cdot N_t \cos \varphi_t\} = 0.$$

The output SNR is therefore

$$(\text{SNR})_{\text{out}} = \frac{P^2}{\sigma^2} = 2\,(\text{SNR})_{\text{in}}.$$

This, of course, is a well-known result.

7.3.2. Weak Signal, Strong Background

The case of a weak signal—which we continue to denote s_t—in a strong background—denoted n_t—which again may be an ensemble of signals and/or noise is mainly of passing interest, since it does not occur in normal operation of the synthetic aperture radar system.‡ By the symmetry of Eq. (7.13), we

† The properties are true for any process for which the integrals exist. More to the point, we did not need any assumptions such as "Gaussian process."

‡ That is, hard limiting normally does not occur, the level being set above the strong background's level.

have immediately that

$$z(t) \approx \frac{4L}{\pi}\cos(\omega_o t + \varphi_t) + \frac{4L}{\pi}\frac{P_t}{2N_t}\cos(\omega_o t + \theta_t) + \frac{4L}{\pi}\frac{P_t}{2N_t}\cos(\omega_o t + 2\varphi_t - \theta_t). \tag{7.14}$$

The result, Eq. (7.14), is perplexing, as we see when we try to "identify" one of the terms as the "output signal." Although the middle term emerges as the best candidate because its phase is correct, its amplitude is rather violently changed by the factor N_t^{-1}: the Fourier transform of P_t/N_t would generally little resemble the impulse response of the system, giving an apparent image much different than that due to a point scatterer. If N_t had a mean m which was large relative to its fluctuation, then $N_t^{-1} \approx m^{-1}(1 - \xi/m)$, and we could then easily identify a signal term: unfortunately, this is not true for the common situation of n_t a Gaussian random process.

This phenomenon, an apparent loss of signal in the output when the signal becomes much weaker than the background ensemble, is known as the "threshold effect."

If the Gaussian background is noise of mean zero and variance σ^2, or if the background is an ensemble of signals and/or noise which approximates such a Gaussian process—as is true for synthetic aperture radar signals—then, as N_t is a Rayleigh random variable, independent of the uniformly distributed random variable φ_t, it is easy to calculate the mean or average output:

$$E\{z_t\} \approx -\left(\frac{2}{\pi}\right)^{1/2}\frac{L}{\sigma}P_t\cos(\omega_o t + \theta_t).$$

Thus, on the average, the output signal is, except for scale, a replica of the input signal.

If in Eq. (7.14) we simply replace N_t^{-1} by its mean value and ignore its variability, then we find the output SNR by simple calculation:

$$(\mathrm{SNR})_{\mathrm{out}} = \frac{\pi}{4}\frac{P_t^2}{2\sigma^2} = \frac{\pi}{4}(\mathrm{SNR})_{\mathrm{in}}.$$

This also is a well-known result. Whether or not the result is meaningful depends upon subsequent processing.

7.3.3. Two Strong Signals

There are times, of course, when multiple strong targets occur, and these can produce false images, as is well known. To see what may happen in such a situation, let the signals be, for simplicity, of constant amplitude,

$$s_1(t) = P\cos(\omega_o t + \theta_t), \qquad s_2(t) = Q\cos(\omega_o t + \alpha_t),$$

and assume that $P \gg Q$. In the case of hard limiting, we have immediately from Eq. (7.17):

$$z_t \approx \frac{4L}{\pi}\left[\cos(\omega_o t + \theta_t) - \frac{Q}{2P}\cos(\omega_o t + \alpha_t) + \frac{Q}{2P}\cos(\omega_o t + 2\theta_t - \alpha_t)\right].$$

The first term is the output due to the strong signal, its level being controlled by the limiter. The second term is the output due to the less strong signal, suppressed by 6 db, by passage through the limiter. The third term is due to the intermodulation of the two signals and can produce false images.

Suppose, for example, that s_1 and s_2 have linear FM, but s_2 is delayed τ sec with respect to s_1 (Fig. 7.3a). In linear operation, both signals compress to images separated by τ sec of course. In nonlinear operation, however, the

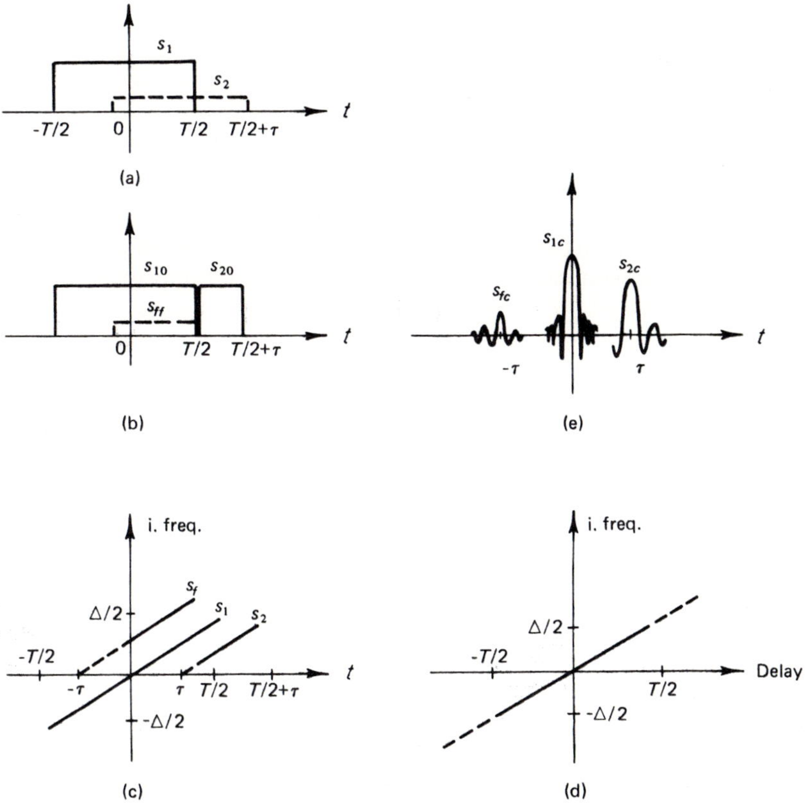

FIG. 7.3 Effect of hard limiting in the compressed image for two high SNR signals, one strong relative to the other.

part of s_2 that overlaps s_1 is suppressed, and the intermodulation term is created during that time (Fig. 7.3b). Using the techniques described in Chapter II for the approximate analysis of linear FM waveforms, it is easy to construct Fig. 7.3e to describe the output. We assume that τ is not too small relative to T, so that the suppressed part of s_2 can be ignored—it compresses to the same location as the unsuppressed part of s_2.

As indicated in Fig. 7.3d, the false target will not be present if $|\tau| > \frac{1}{2}T$, if the compression filter is sharply cut off outisde the spectrum of the signals. In practice, this means that the false target s_{fc} will be somewhat reduced in amplitude. We also note that the less strong signal is reduced in resolution and amplitude in direct proportion to the amount of overlap for, again, τ not too small relative to T.

When τ goes to zero, the suppressed part of s_2 dominates the unsuppressed part, of course: then the amplitude of the less strong signal's compressed output becomes proportional to P/Q, as does the false image, and both approach full resolution.

7.4. RANDOM SIGNALS AND RANDOM NOISE

The methods of Sec. 7.3 are useful to describe and predict anomalous behavior due to strong signals. The method is less satisfactory in analyzing the effects on weak signals. Obviously, the method gets tedious and then useless, as the number of strong signals increases. To describe how the majority— as distinct from the anomalous—signals are affected, we approximate the ensemble of signals by a Gaussian process and use a mean-square error criterion. We again assume additive, Gaussian noise and an error function limiter followed by a filter. Unlike before, we shall be able to obtain results in the low pass case.

We shall generalize the model in order to prevent unrealistic solutions by adding a secondary noise source—e.g., film noises in the recording operation. The model is shown in Fig. 7.4. We minimize the mean-square error:

$$\min_{K,H} \sigma_\varepsilon^2(K, H) \equiv E(z - s)^2,$$

where K is the prefilter gain, and H is the post filter.

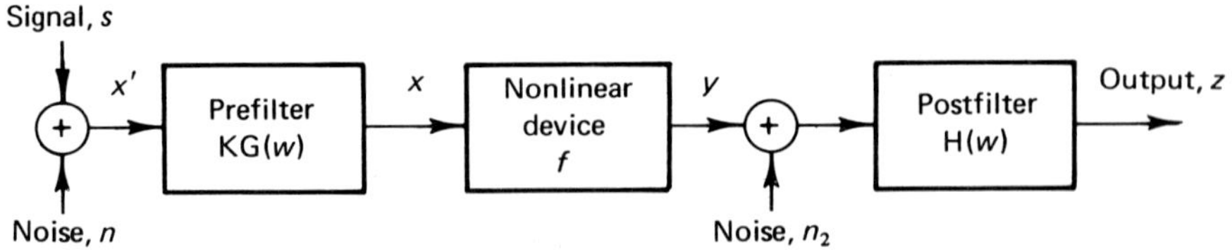

FIG. 7.4 System nonlinear model with secondary noise source.

Noting that the error process is $\varepsilon = z - s$, one finds its covariance function and hence its spectral density $S_\varepsilon(\omega)$ which may be written in terms of the spectral densities of the other processes as

$$S_\varepsilon(\omega) = \left| [S_{f(x)}(\omega) + S_2(\omega)]^{1/2} H(\omega) - \frac{S_{s,\,f(x)}(\omega)}{[S_{f(x)}(\omega) + S_2(\omega)]^{1/2}} \right|^2$$

$$+ S_s(\omega) - \frac{|S_{s,\,f(x)}(\omega)|^2}{S_{f(x)}(\omega) + S_2(\omega)}, \tag{7.15}$$

where $S_{s,\,f(x)}(\omega)$ is the cross-spectral density of s and $f(x)$.

Allowing for nonrealizable filters (which are approximable by realizable filters with delay) and noting that only the first term depends on $H(\omega)$, one has the optimum postfilter

$$H(\omega) = \frac{S_{s,\,f(x)}(\omega)}{S_{f(x)}(\omega) + S_2(\omega)}. \tag{7.16}$$

The residual mean-square error that is to be minimized by the prefilter gain K is then

$$\sigma_\varepsilon^2 = \frac{1}{2\pi} \int \left[S_s(\omega) - \frac{|S_{s,\,f(x)}(\omega)|^2}{S_{f(x)}(\omega) + S_2(\omega)} \right] d\omega. \tag{7.17}$$

Equivalently, one chooses K to maximize the form

$$\Omega(K) = \int \frac{|S_{s,\,f(x)}(\omega)|^2}{S_{f(x)}(\omega) + S_2(\omega)} \, d\omega. \tag{7.18}$$

Of course, one must find the rather complicated spectra appearing in Eqs. (7.16) and (7.18). They can be determined by the characteristic function method. In the event that $\{s_t,\ t \in T\}$ and $\{x_t,\ t \in T\}$ are jointly Gaussian processes, one may couple the following two facts from statistics:

$$E[s_{t+\tau} f(x_t)] = E[f(x_t)\, E(s_{t+\tau} \mid x_t)]$$

and

$$E(s_{t+\tau} \mid x_t) = \sigma_x^{-2} R_{s,\,x}(\tau) x_t,$$

to note

$$R_{s,\,f(x)}(\tau) = \sigma_x^{-2} E[f(x_t) x_t] R_{s,\,x}(\tau). \tag{7.19}$$

We assume that the nonlinear device f is characterized by

$$f(x) = \frac{L}{\theta} \left(\frac{2}{\pi}\right)^{1/2} \int_0^x \exp\left[-\frac{1}{2}\frac{\lambda^2}{\theta^2}\right] d\lambda.$$

It is known that, upon application of the characteristic function method, one obtains

$$R_{f(x)}(\tau) = \frac{2L^2}{\pi} \arcsin\left(\frac{R_x(\tau)}{R_x(0) + \theta^2}\right).$$

Assuming that the input signal is an independent Gaussian process and using (7.19), one has

$$R_{s_o,\, f(x)}(\tau) = \left(\frac{2}{\pi}\right)^{1/2} \frac{L}{(\theta^2 + \sigma_x{}^2)^{1/2}} R_{so}(\tau),$$

where $\{s_{o_t},\, t \in T\}$ is the signal process at the input of the nonlinearity.

The prefilter $KG(\omega)$ is most easily accounted for by letting it be of gain K and lumping the effect of $G(\omega)$ into $\sigma_x{}^2$ in the usual way. Then, for the system being considered, one has

$$R_{s,\, f(x)}(\tau) = \left(\frac{2}{\pi}\right)^{1/2} \frac{L}{\left(\sigma_x{}^2 + \dfrac{\theta^2}{K^2}\right)^{1/2}} R_s(\tau),$$

and

$$R_{f(x)}(\tau) = \frac{2L^2}{\pi} \arcsin \frac{R_x(\tau)}{\sigma_x{}^2 + \dfrac{\theta^2}{K^2}}. \tag{7.20}$$

It is a good approximation to expand the arcsin in its Taylor series and retain only the first two terms, because the spectral contributions of the higher-order terms are of rapidly decreasing magnitude and are spread over greater bandwidths [and hence can be filtered out by $H(\omega)$].

Substituting these results into Eq. (7.20), one has

$$\Omega(K) = \int S_s{}^2(\omega)\, d\omega / [S_s(\omega) + S_n(\omega) + (\sigma_x{}^2 + \theta^2 K^{-2})^{-2} S_3(\omega)$$

$$+ \frac{\pi}{2L^2} (\sigma_x{}^2 + \theta^2 K^{-2}) S_2(\omega)], \tag{7.21}$$

where the third-order spectrum is

$$S_3(\omega) = {}_3S_3(\omega) + 3\ {}_{2,\,1}S_{s,\,n}(\omega) + 3\ {}_{1,\,2}S_{s,\,n}(\omega) + {}_3S_n(\omega)$$

and

$${}_{n,\,m}S_{u,\,v}(\omega) \equiv \mathscr{F}\{R_u{}^n(\tau) R_v{}^m(\tau)\},$$

where $\mathscr{F}$ denotes the Fourier transform.

It is not possible, of course, to determine K without specifying the three input spectral densities. However, two limiting cases are clear. First, if

$S_2(\omega) = 0$, then $\Omega(K)$ is maximized when $K = 0$. (In this case, $H(\omega)$ is required to furnish infinite gain; the practical limitations of this solution are evident.) Second, if the third-order spectra $S_3(\omega)$ are negligible compared to $S_2(\omega)$ for all ω in the spectrum of $\{s_t, t \in T\}$, then $K = \infty$ is optimum. The compromise between minimizing the distortion introduced by operation in the nonlinear region and the necessity of overcoming the secondary noise source is clearly indicated.

As a specific choice of spectra, in some signal spectral band Δ_s of bandwidth $\Delta\omega$ centered on $\pm\omega_c$, let

$$S_s(\omega) = A, \qquad S_n(\omega) = N, \qquad S_2(\omega) = N_2 .$$

Then $G(\omega)$ will be taken as an ideal filter with band pass Δ_s, assuming $S_s(\omega) = 0$, $\omega \notin \Delta_s$. It is assumed that Δ_s has bandwidth sufficiently narrow so that the third-order spectra do not overlap significantly.

Then, one can compute that

$$S_3(\omega) = \frac{3(A + N)^3}{4\pi^2}\left[-(\omega - \omega_c)^2 + \frac{3\Delta\omega^2}{4}\right], \qquad \omega \in \Delta_s .$$

Since it is readily seen that $S_3^{\max}/(A + N) = 3/32$, and that $S_3(\omega)$ varies over Δ_s by only one third of its value at ω_c, it is a good approximation to assume that

$$S_3(\omega) \approx S_3(\omega_c) = \frac{9}{16\pi^2}(A + N)^3\,\Delta\omega^2, \qquad \omega \in \Delta_s .$$

Substituting these results into Eq. (7.17), one has for the mean-square error

$$\sigma_\varepsilon^{\,2} = \sigma_s^{\,2}\left\{1 - \frac{1}{(1 + 1/\psi_i)[1 + \frac{3}{32}(1 + \overline{K}^{-2})^{-2} + (1 + \overline{K}^{-2})\bar{\sigma}_2^{\,2}]}\right\}, \qquad (7.22)$$

where $\overline{K}^2 = K^2\sigma_x^{\,2}/\theta^2$, $\bar{\sigma}_2^{\,2} = \pi\sigma_2^{\,2}/2L^2$, and ψ_i is the SNR at the input of the nonlinear device. Equation (7.22) is plotted in Fig. 7.5. It is seen that $K = \infty$ is the optimum solution, if the (normalized) variance of the secondary noise source is greater than $\frac{3}{16}$. However, the decrease in $\sigma_\varepsilon^{\,2}$ at the finite minimum when $\bar{\sigma}_2^{\,2} < \frac{3}{16}$ can be shown [using (Eq. 7.22)] to be less than 10%.† *If we assume a low-pass spectrum* ($\omega_o \equiv 0$), *then the result is essentially the same: the coefficient* $\frac{3}{32}$ *becomes* $\frac{4}{32}$.

† Most significantly,

$$\lim_{\{\psi_i \to \infty,\, \sigma_2^2 \to 0\}} [\sigma_\varepsilon^{\,2}] = \frac{3}{35}\sigma_s^{\,2},$$

all due to distortion.

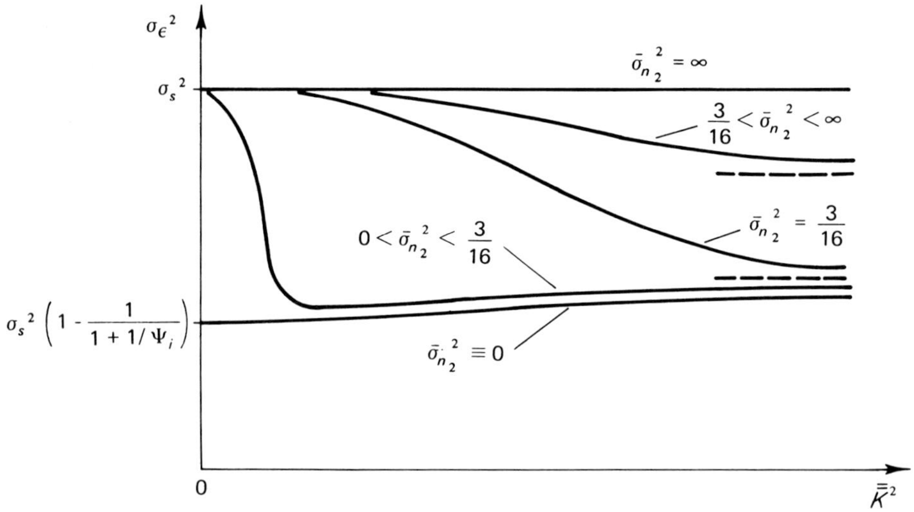

FIG. 7.5 The mean square error $\sigma_\varepsilon^2 = E\{(Z_t - S_t)^2\}$ as a function of the normalized prefilter gain, for fixed input SNR.

Since the distortion spectra are so small, the postfilter $H(\omega)$ is nearly the usual Wiener infinite-lag filter, reflecting the fact that the behavior of the nonlinear system being considered can be essentially linear in mean-square properties. From Eq. (7.16) with $K = \infty$, one has the optimum postfilter

$$H(\omega) = \begin{cases} 4(2\pi)^{1/2} \dfrac{\sigma_s^2}{L\sigma_x} \dfrac{1}{\dfrac{35}{4} + 8\bar{\sigma}_2^{\,2} - \left(\dfrac{\omega - \omega_c}{\Delta\omega}\right)^2}, & \omega \in \Delta_s, \\ 0, \quad \omega \notin \Delta_s. \end{cases}$$

Thus, as might be expected, $H(\omega)$ favors slightly the signal frequencies near the edge of the signal band. It is noteworthy that the behavior of this class of nonlinear devices is predominantly linear with respect to mean-square properties.

Some applications of the results. An important fact concerning the prefilter is apparent from Eqs. (7.16) and (7.17); if the input processes x and s are jointly Gaussian, then only the amplitude characteristic of $G(\omega)$ is significant. Hence, the optimum filter $H(\omega)$ and the optimum prefilter gain K are independent of the phase characteristic of the prefilter $G(\omega)$. For example, this allows uncompromised freedom to use the phase characteristic in coherent processing of point target information by pulse compression systems.

COMMENTS AND REFERENCES

The characteristic function method is discussed in standard texts; see, e.g., Davenport and Root (1958). There is a continuing sequence of papers on no-memory nonlinearities with random inputs: Galejis (1959), Cahn (1961), Jones (1963), and Berglund (1964).

For the required forms from classical analysis such as the Jacobi-Anger formula, the integral of Weber and Sonine, etc., see Magnus and Oberhettinger (1949).

Nonlinear devices with memory are much less susceptible to general conclusions. For a discussion of a model that can represent negative resistance amplifiers that are in use in some SAR systems, see Harger and Lewinski (1967).

The model we have used in this chapter is quite useful in modeling photographic film and other nonlinearities; see Kelly (1960).

VIII

Optimum Ambiguity Design

8.1. INTRODUCTION

We shall now take up the problem of optimum design of the SAR system when the desired range interval is of the order of the unambiguous range interval. That is, we will be concerned in this chapter with the design of the SAR system when operating at the ambiguous limit of range interval. Such a situation is commonly encountered when the radar set is aboard an orbital vehicle.

We recall that the problem of ambiguities arose because the physical antenna necessarily has a finite aperture, and as a consequence the far-field pattern of the antenna cannot be confined solely to the desired solid angle. This, in turn, is due to the fact that a function and its Fourier transform cannot both be zero, outside a finite interval. Additionally, there is an inherent dependence between parameters that we would like to specify independently—principally, as we saw, unambiguous range interval and azimuth resolution [Eq. (1.1)].

Among the SAR system parameters to be determined by an optimum design is the PRF. As the PRF is varied, the angular intervals where the antenna pattern is desirably contained also vary and, therefore, also the desirable antenna illumination functions. Of course, it is the sampling operation that gives rise to the ambiguities as we saw (Sec. 3.4): the aliases that are formed have segments in their sidelobe region that end up in the spectral band of the processor—e.g., matched filter—and cause spurious images.

We shall assume that the processor is the one we found optimum under both the SNR_{max} and ML criteria (Chapter IV), and whose realization by a coherent optical system we discussed in Chapter V: the so-called "matched filter" processor. We have already derived the complete impulse response of this system in Sec. 3.4 and observed the presence of ambiguities. The matched filter system's impulse response we shall call the *ambiguity function* (see Fig. 8.1).

We shall see when we examine the spurious responses in the SAR matched filter system, that the azimuth ambiguities strongly suggest a certain energy criterion for the design of the azimuth antenna pattern, and that a certain average antenna gain criterion is very reasonable for the design of the vertical

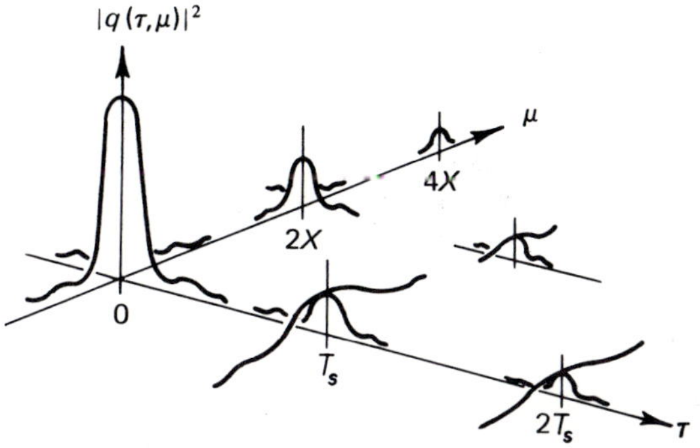

FIG. 8.1 Typical ambiguity function for a side-looking radar. Azimuth offset operation.

antenna pattern. We shall let the sidelobe behavior of the central peak of the ambiguity function fend for itself—quite unlike the approach described in Chapter IV for "unambiguous" systems.

We can place criteria on the central response of the ambiguity function in the range dimension because, as we have already observed (Sec. 3.4), the ambiguity function essentially factors when a periodic modulation is used. Thus, we shall proceed as in Chapter IV and will, in fact, use an energy criterion as discussed in Sec. 4.2.1.

We shall not, in this chapter, discuss attempts to avoid or suppress ambiguities to overcome, e.g., the "fundamental" limitation on range interval. These matters will be taken up in Chapter IX.

8.2. AMBIGUITY FUNCTION DESIGN CRITERIA

Azimuth criterion. We shall begin the discussion of appropriate criteria for SAR system design by assuming that, in fact, the LIV processor filter has matched phase but uniform amplitude, and that range offset is being employed. This will be a somewhat simpler beginning, and we shall, of course, extend the discussion.

Referring to Sec. 3.4, as we will be doing frequently, and specifically to

Eqs. (3.14)–(3.20)—and mentally modifying them, since the processor filter has uniform amplitude—we see that:

1. The height of the ambiguous peaks in azimuth is proportional to the energy of the one-way voltage pattern in azimuth ("VPA") in the angular sectors centered on $n\ X/R$, $n = \pm 1, \pm 2, \pm \ldots$, and of width X/R—these sectors are termed "ambiguous azimuth intervals."
2. The height of the central response of the ambiguity function is proportional to the energy of the VPA in the "unambiguous" azimuth angular interval $(-X/2R, +X/2R)$.

Thus, we obviously should maximize the energy of the VPA in the unambiguous interval relative to the energy in any particular ambiguous interval. Since a scattering object is presumably equally likely to occur in any ambiguous interval, we decide that:

The design criterion for the one-way voltage pattern in azimuth is to maximize the energy in the unambiguous angular interval.

If the SAR system employs an azimuth offset (vice a range offset), then we have noted [Eq. (3.18)] that the contribution to the ambiguity function of the odd numbered ambiguous intervals is a dispersed—vice compressed— response. This, of course, was because the phase histories of such segments have linear FM slope of incorrect sign. Experience has shown that if the time-bandwidth-product is large—as it would be here—such spurious responses do not noticeably detract from image "appearance," and, hence, one might be tempted to allow energy of the VPA to fall into these odd intervals in order to control the VPA better in the even-numbered intervals—which do result in compressed spurious responses. However, partial results in this direction indicate that, in cases of practical interest, such a criterion leads to essentially the same patterns as the originally proposed criterion.† We therefore retain the same criterion for both azimuth and range offset operation SAR systems.

Next we examine the situation when the processor is a matched filter—that is, the processor is matched in both amplitude and phase of the signal. The peaks of the azimuth ambiguities are not now so simply related to the VPA. However, as we shall see, the amplitude of the azimuth signals do not differ greatly from uniformity. The taper will be found to increase with increasing "freedom" in the design parameters. However, the approximation that the peaks of the azimuth ambiguities is proportional to the energy of the VPA in the ambiguous intervals still seems to be good. We, therefore, continue to retain the same criterion for design of the VPA.‡

† See Appendix 8A.

‡ Appendix 8C considers this point in more detail.

We point out again that we are not directly controlling the sidelobe level in the azimuth direction of the central response of the ambiguity function—obviously, since there is no mention of sidelobe levels in our criterion. To place such an additional criterion would almost certainly lead to conflicting demands on the VPA, since it determines the entire azimuth behavior, of course. However, we shall find that the sidelobe levels that do, in fact, result are quite compatible with the resulting ambiguity levels (Figs. 8.8 and 8.9).

Range ambiguity criterion. From our derivation (in Sec. 3.4) of the ambiguity function for the SAR system—and as is also obvious, anyway—we see that the two-way voltage pattern of the antenna in the vertical plane is a multiplicative factor dependent on range. Thus, the gain—defined as the square magnitude of the voltage pattern—at any location in the unambiguous range interval should be as large as possible, relative to the gain assigned to any ambiguous location. The following choice is therefore reasonable:

> The design criterion for the vertical antenna pattern is to maximize the average gain in the unambiguous angular interval.

Of course, maximizing the average gain is the same as maximizing the relative energy in said interval: we, therefore, have arrived at the same criterion for the horizontal (azimuth) and vertical (range) one-way voltage patterns of the physical antenna.

In deciding upon this criterion, we have not directly concerned ourselves with (1) the uniformity of gain over the unambiguous range interval—rather obviously, a generally desirable property—or (2) the inherent suppression of range ambiguities, especially in typical orbital geometries. In fact, the latter consideration, (2) allows a relaxing, roughly speaking, of the requirements on the gain pattern, thereby meeting the requirements of the first consideration, (1), more satisfactorily. Thus, we could formulate a criterion that emphasizes uniform gain; in fact, such a solution would logically be restrained to be linear summations of prolate spheroidal wave functions, since they form a complete basis for the aperture interval. However, we will not pursue the matter here: we consider the solutions we do obtain to be acceptable with respect to gain uniformity, especially in orbital SAR systems.

Range sidelobe criterion. We have already pointed out that sidelobes cause spurious responses of the same order of seriousness as ambiguities, and that the range sidelobes of the ambiguity function are entirely determined by the system's signal modulation. We have already discussed this problem in Sec. 4.2, and we adopt a criterion whose merits were discussed there:

> The design criterion for the signal modulation is to maximize the

energy of the range cross section, $|q(\tau, o)|$, of the ambiguity function in the range resolution interval.

Summary of criteria. We have, therefore, arrived at—not entirely innocently!—the same criterion to determine the azimuth antenna pattern, the vertical antenna pattern, and the signal modulation. The mathematical problem in all three cases is the same: we want to find a function that when truncated—e.g., placed in an aperture—then Fourier transformed—e.g., propagated to the far-field—and then truncated again—e.g., examined only over the unambiguous interval—has the maximum possible relative energy. For the sake of clarity in the details, we shall first consider the antenna-design problems and then the signal-modulation-design problems. In fact, the latter problem has further complications that we will have to resolve.

8.3. SOLUTION OF THE ANTENNA DESIGN PROBLEM

From the foregoing, we see that the mathematical statement of the antenna design problem is

$$\max_f \frac{\int_I b(u)\,du}{\int_{-\infty}^{\infty} b(u)\,du},$$

where in the horizontal (azimuth) pattern problem, $b = |B_h|$, and

$$I = I_h \equiv \frac{X}{R_o}, \tag{8.1}$$

and in the vertical (range) pattern problem $b = |B_v|$, and

$$I = I_v \equiv \frac{v \tan \gamma_o}{2R_o f_s}. \tag{8.2}$$

We have assumed I_h and I_v are small angles (Fig. 8.2).

The far-field pattern magnitudes $|B_h|$ and $|B_v|$ are given by Eqs. (3.8a) and (3.8b), respectively. (Here we are using angular coordinates.) Thus, e.g.,

$$|B_h(u)| = \left| \int_{-\Omega_h}^{\Omega_h} e^{-ikux} f(x)\,dx \right|^2, \tag{8.3}$$

where $(-\Omega_h, \Omega_h)$ is the physical antenna's azimuth aperture, $f(x) \equiv E_{oh}(x)$, $x \in (-\Omega_h, \Omega_h)$, and $k \equiv 2\pi/\lambda$.

Using Eq. (8.3), it is eay to see that, if we define

$$S_o(\omega) \equiv f(\Omega\omega) \tag{8.4}$$

and

$$c \equiv \tfrac{1}{2}kI\Omega, \tag{8.5}$$

our problem is to

$$\max_{S_o \in \mathscr{L}^2} \frac{(AS_o, S_o)_1}{\|S_o\|_1^{\,2}}, \tag{8.6}$$

where the operator A is such that

$$Af(\omega) \equiv \int_{-1}^{+1} \frac{\sin c(\omega - v)}{\pi(\omega - v)} f(v)\, dv.$$

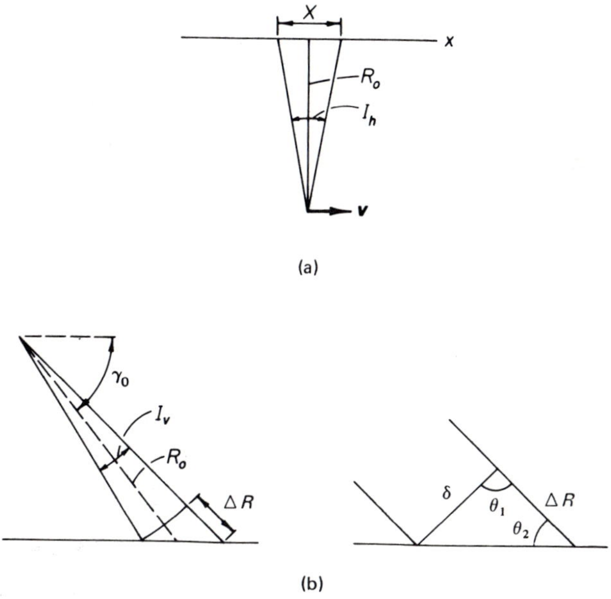

FIG. 8.2 Geometry of SAR and small angle assumptions. Curve (a) Slant range ("horizontal") plane: $I_h \approx X/R_0$. Curve (b) Vertical plane: $\theta_1 \approx \frac{1}{2}\pi$, $\theta_z \approx \gamma_o$, $I_v \approx \delta/R_o$.

The notations $(\cdot\,,\cdot)_1$ and $\|\cdot\|_1$ are the inner product and norm on $\mathscr{L}^2\{[-1, 1]\}$, the space of all functions of finite (Lebesque) integral-square over the interval $[-1, 1]$,† We note that I and Ω enter only as their product.

† In order to have a well-defined mathematical problem, we must choose some collection (space) of functions over which we search in order to maximize the functional in Eq. (8.6). $\mathscr{L}^2\{[-1, 1]\}$ is chosen because of its well-developed theory and its "reasonableness" for modeling practical problems.

Since A is a (linear) completely continuous, symmetric operator, it is well known that a unique solution to this problem exists, and that the value of the maximum is given by the largest positive characteristic value, μ_o^+, of the integral equation

$$AS_o = \mu S_o. \tag{8.7}$$

The associated characteristic function, S_{oo}, is the sought-for solution.

The characteristic functions $\{S_{on}\}$ of the integral Eq. (8.7) turn out to be prolate spheroidal wave functions (PSWF's), and a very thorough, elegant, and useful investigation of them has been made by Slepian and Pollack (1961). For our purposes, a tabulation has been made and is shown in Figs. 8.3 and 8.4.

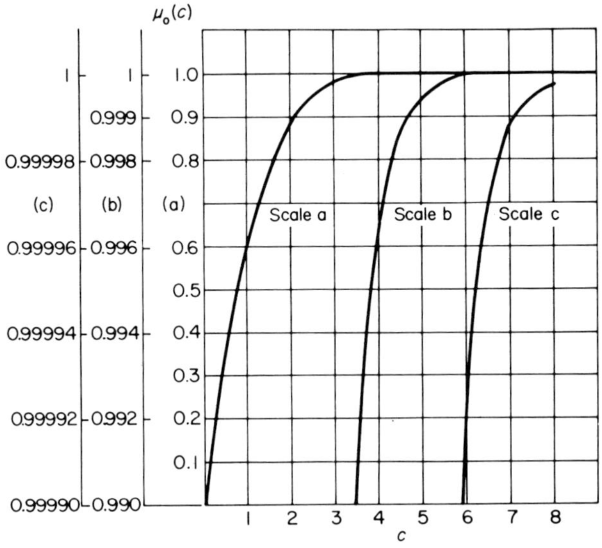

FIG. 8.3 The largest characteristic value $\mu_o(c)$. This is the fraction of total energy in the unambiguous intervals for the antenna patterns or in the resolution interval for the delay cross section of the ambiguity function.

Figure 8.3 gives the value of the largest characteristic value, $\mu_o(c)$, as a function of c. We note that a seemingly very large fraction of the pattern's energy can be contained in the desired intervals for "modest" (as we will make clear below) values of c. The associated characteristic functions $\{S_{oo}(c; \cdot)\}$, are shown in Fig. 8.4 for the range of c of interest here; they are a tapered family of aperture illuminations that are, roughly at least, reasonable aperture illuminations to realize in practice.

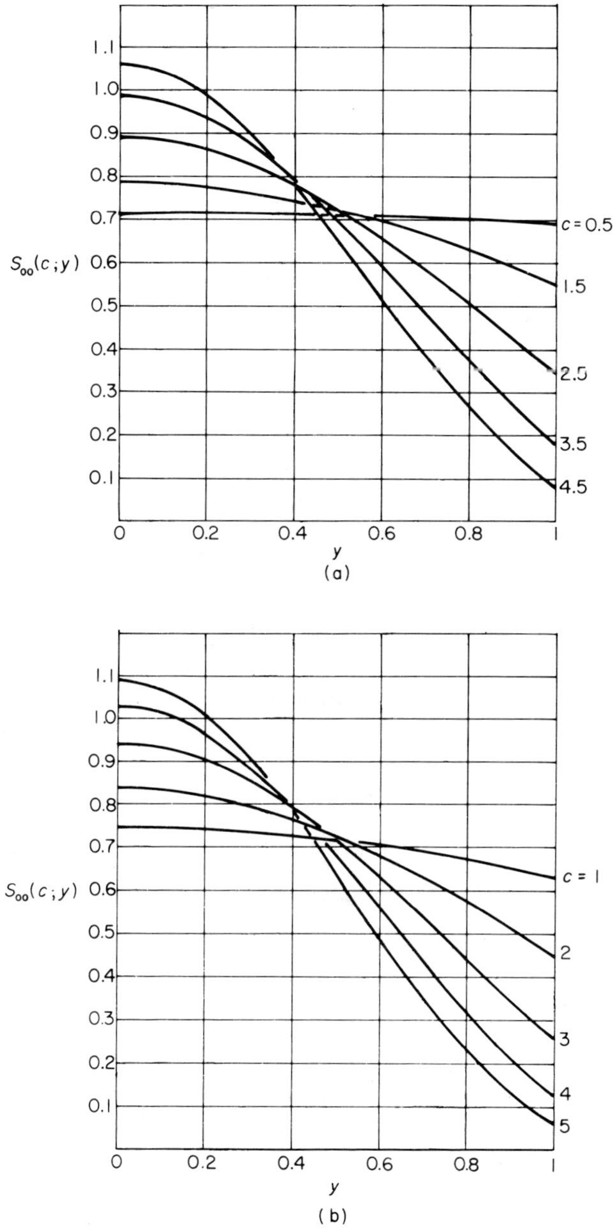

FIG. 8.4 The characteristic function $S_{oo}(c; y)$, associated with the largest characteristic value. These are the optimum antenna aperture illumination functions (setting $y = x/\Omega_h$, or $y = x/\Omega_v$), or the optimum signal spectrum square magnitude (setting $y = \omega/\Omega_s$).

As pointed out by Slepian and Pollock (1961), the prolate spheroidal wave functions reproduce themselves (except for scale) when truncated to $[-1, 1]$ and then Fourier transformed. Therefore, we can solve the integral Eq. (8.7) for points outside $[-1, 1]$ and thereby determine the Fourier transform over a larger interval. This continuation is shown in Fig. 8.5.

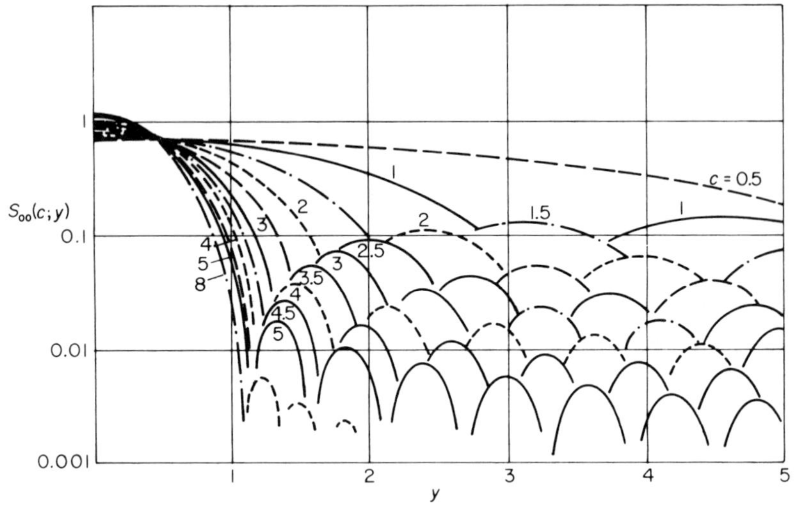

FIG. 8.5 Continuation of the characteristic function $S_{oo}(c; y)$. These are the optimum antenna far-field one-way voltage patterns (setting $y = 2\mu/I_h$, or $y = 2\mu/I_v$), or the optimum delay cross section of the ambiguity function (setting $y = 2\tau/I_s$).

We can now calculate the azimuth cross section $|q(o, \mu)|$ in the unambiguous interval: it is easy to show that

$$|q(o, \mu)|^2 = \left| \int_{-1}^{+1} e^{i\pi x u} |S_{oo}(c_h; u)|^2 \, du \right|^2. \tag{8.8}$$

This output response is given in Figs. 8.6. and 8.7

From these results, we can note the relative heights (average relative height in the range dimension) of the major spurious responses of the ambiguity function. These are summarized in Figs. 8.8, 8.9, and 8.10. Curve (b) of Fig. 8.10 takes into account a typical value of the inherent suppression of range ambiguities due to the SAR geometry that occur in orbital applications; a fairly good estimate of this effect can be obtained from a routine calculation similar to those made in Sec. 2.6.

Since the width of the impulse response between its half-power points is a common resolution measure, this is also plotted in Fig. 8.11 from the data

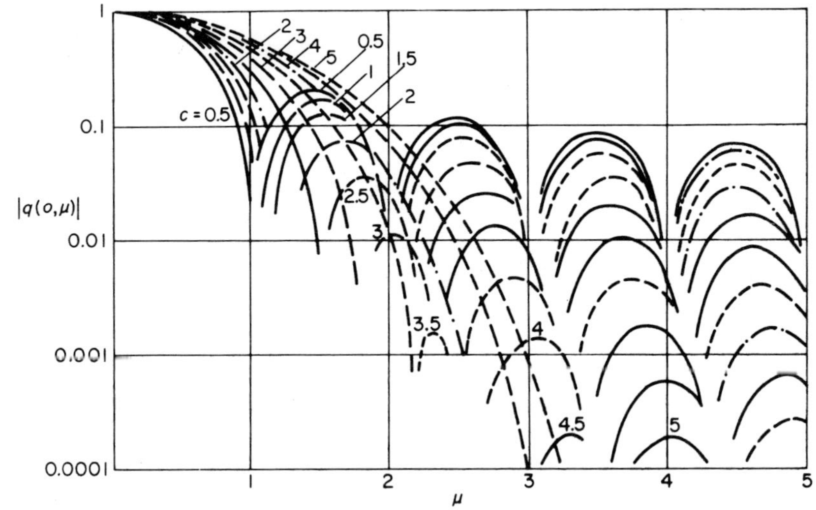

FIG. 8.6 The azimuth cross section of the ambiguity function for phase-matched processing.

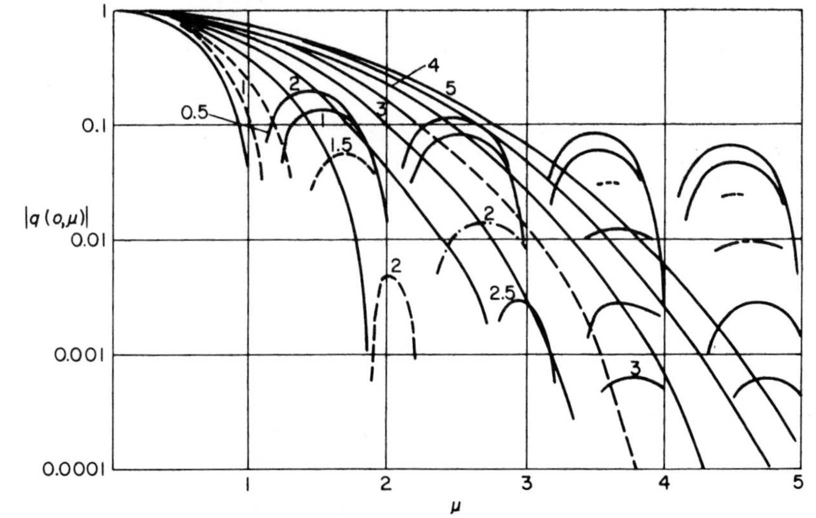

FIG. 8.7 The azimuth cross section of the ambiguity function for matched processing.

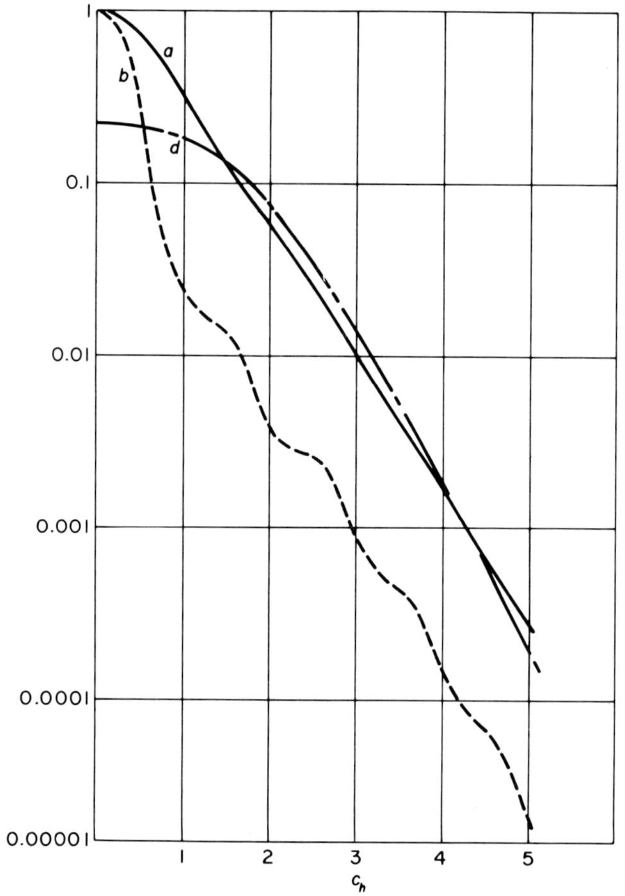

FIG. 8.8 Important characteristics of azimuth cross section $|q(o, \mu)|$ for phase-matched processing. Curve (a) relative height of the first ambiguous peak, range offset. Curve (b) relative height of the first ambiguous peak, azimuth offset. Curve (d) relative height of the largest sidelobe.

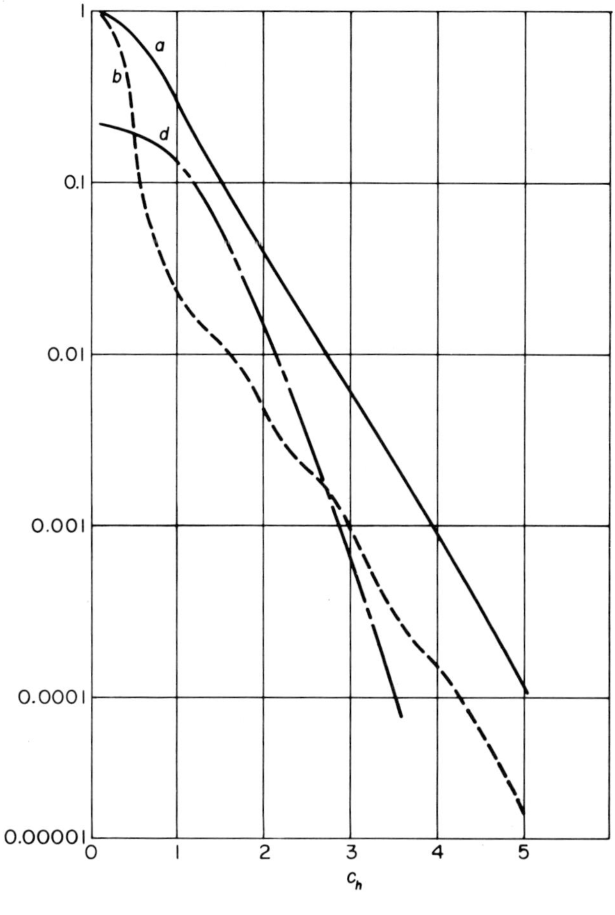

FIG. 8.9 Important characteristics of azimuth cross section $|q(o, \mu)|$ for matched processing. Curve (a) relative height of the first ambiguous peak, range offset. Curve (b) relative height of the first ambiguous peak, azimuth offset. Curve (d) relative height of the largest sidelobe.

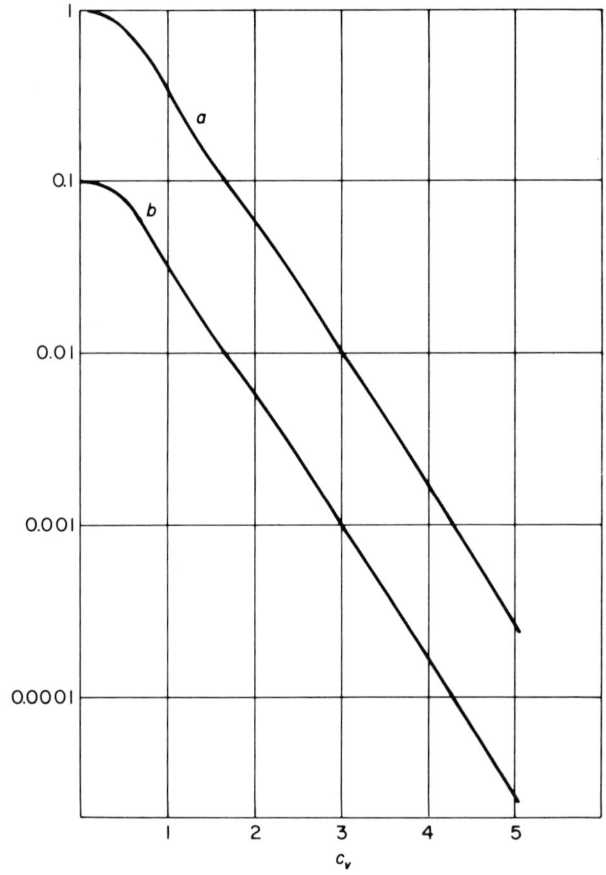

FIG. 8.10 Important characteristics of range cross section. Relative average height of largest ambiguous peak for inherent suppression of 0 db [curve (a)] and 10 db [curve (b)].

presented in Figs. 8.6 and 8.7. Of course, a similar plot could be made for similar resolution measures, using Figs. 8.6 and 8.7.

We observe that the azimuth sidelobe levels—which are not directly controlled by the criterion we adopted—are quite satisfactory in that they roughly follow the ambiguity levels—see Figs. 8.8 and 8.9. In fact, for an SAR system using phase-matched processing and range offset, we have an extremely good design: curves (a) and (d) of Fig. 8.8.

Examining curve (b) of Fig. 8.10 we see that very low (average) range ambiguity levels are obtained with relatively small values of c_v: referring additionally to Fig. 8.4, we see that the gain pattern over the unambiguous range interval will be acceptably nonuniform, varying typically by a couple of decibels.

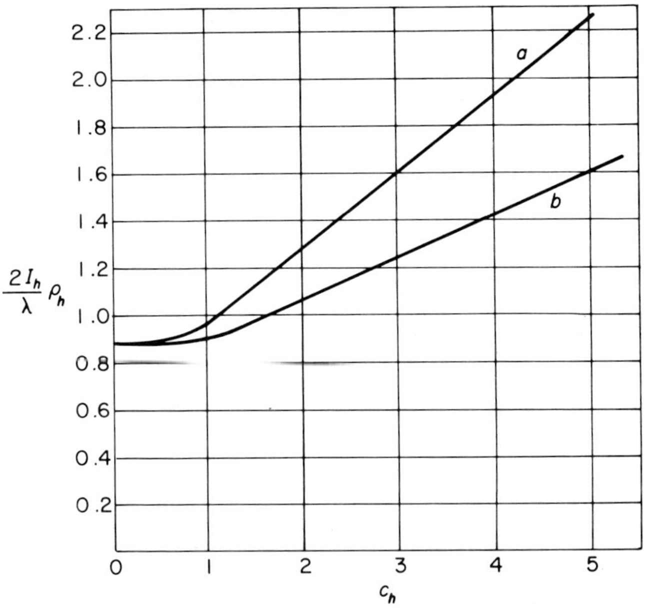

FIG. 8.11 Azimuth resolution ρ_h. Defined as the half-power width of main response of $|q(o, \mu)|$. Curve (a) matched processing; curve (b) phase-matched processing.

8.4. SOLUTION OF THE SIGNAL DESIGN PROBLEM

From the foregoing discussion (Sec. 8.2), we see that the mathematical statement of the signal modulation design problem is

$$\max_{\tilde{F} \in \mathscr{L}^2} \frac{\int_{I_S} |q(\tau, o)|^2 \, d\tau}{\int_{-\infty}^{\infty} |q(\tau, o)|^2 \, d\tau},$$

where I_S is the range, or delay, resolution interval; for phase-matched filtering,

$$q(\tau, o) = \int_{-\Omega_s}^{\Omega_s} e^{i\omega\tau} |\tilde{F}(\omega)| \, d\omega,$$

and for matched filtering,

$$q(\tau, o) = \int_{-\Omega_s} e^{i\omega\tau} |\tilde{F}(\omega)|^2 \, d\omega.$$

The bandwidth of the signal modulation is $2\Omega_s$.

We can, of course, immediately write down the solutions in light of the discussion of Sec. 8.3:

$$|\tilde{F}(\omega)| = \begin{cases} S_{oo}\left(c_S; \dfrac{\omega}{\Omega_S}\right), & |\omega| \le \Omega_S, \quad \text{phase matched,} \\ S_{oo}^{1/2}\left(c_S; \dfrac{\omega}{\Omega_S}\right), & |\omega| \le \Omega_S, \quad \text{matched,} \end{cases}$$

where

$$c_S \equiv \tfrac{1}{2}\Omega_S I_S \tag{8.9}$$

Analogously to Ω and I in the antenna problem, Ω_S, and I_S enter only as their product in determining the functional form of $|\tilde{F}|$.

We have, therefore, determined the magnitude $|\tilde{F}|$ of the complex modulation spectrum, by specifying the percent energy of $|q(\tau, o)|$ that must be contained in the desired resolution interval I_S. Of course, the phase of F remains free as yet. In fact, this is fortunate: as we have already observed, signal energy determines the SNR in matched filter systems, and practical peak power restraints force one to the use of dispersed—that is, phase-modulated—modulations in order to meet both bandwidth and signal energy requirements in some applications.

We would, therefore, like to independently choose the magnitude $|F|$ of the complex modulation to satisfy the signal energy and peak power restraints. It is not a priori clear that this can be done; that is, it is not a priori obvious that there exists a phase modulation θ such that $|F|\, e^{i\theta}$ has spectral magnitude $|\tilde{F}|$. It can be shown† that an approximate solution is possible if, roughly speaking, the pair $|F|$ and $|\tilde{F}|$, are of large time-bandwidth product. The requisite phase modulation is given by Eq. (8B.4) of App. 8B as

$$\theta'(c_S; t) = G_*^{-1}\{2\pi G(t)\},$$

where $\theta' \equiv d\theta/dt$ is the frequency modulation;

$$G(t) \equiv \int^t |F(S)|^2\, dS, \qquad G(\infty) = 1,$$

and G_*^{-1} is the inverse (assumed to exist) of the function

$$G_*(x) \equiv \int^x |\tilde{F}(y)|^2\, dy.$$

We continue by assuming an eminently practical choice of signal envelope: it will be proportional to

$$|F(t)| \equiv T^{-1/2} \operatorname{rect}\left(\frac{t}{T}\right).$$

Again, T will be determined by the SNR requirement and the peak power restraint. The resulting frequency modulations are given in Fig. 8.12 for

† See Appendix 8B.

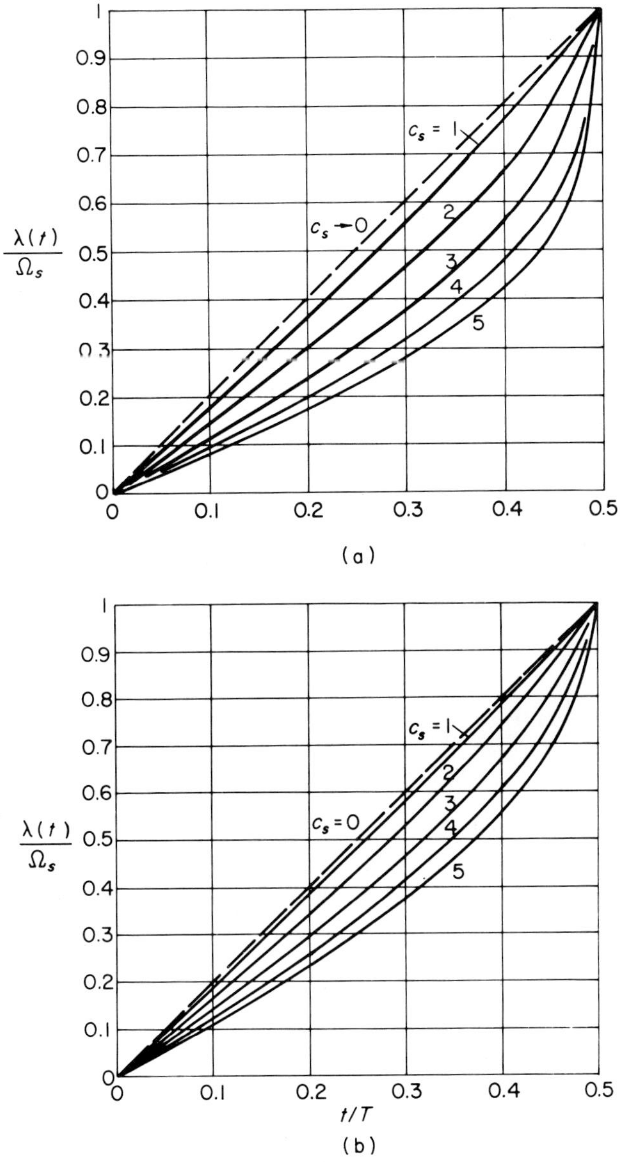

FIG. 8.12 The optimum signal frequency modulation. Curve (a) phase matched processing; curve (b) matched processing.

phase-matched and matched processors. One can generate such modulations quite easily by, e.g., using an oscillator whose frequency can be varied. An approximate, nonoptimum procedure that would probably be acceptable in practice would be to generate a linear FM pulse and perform the above spectral weighting in the processor.

The resulting sidelobe levels are shown in Fig. 8.13. Also of interest is the resultant 3 db resolution: this is shown in Fig. 8.14. The typical tradeoff between such resolution measures and sidelobe level is evident.

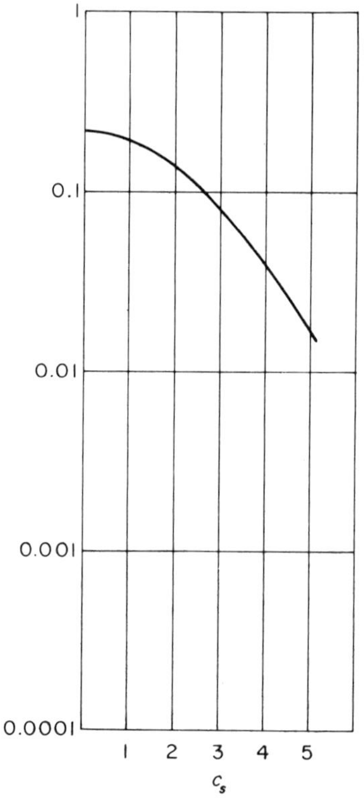

FIG. 8.13 Relative height of largest sidelobe of range cross section.

8.5. EXAMPLE

There are several ways one can use the results we have developed in this chapter to find optimum designs of SAR systems. As a "canonical" type, suppose we have the following problem:

We wish to design an SAR system using a vehicle at a (mean) slant range R_* naut mi, depression angle γ_*, and a velocity v_* ft/sec. The

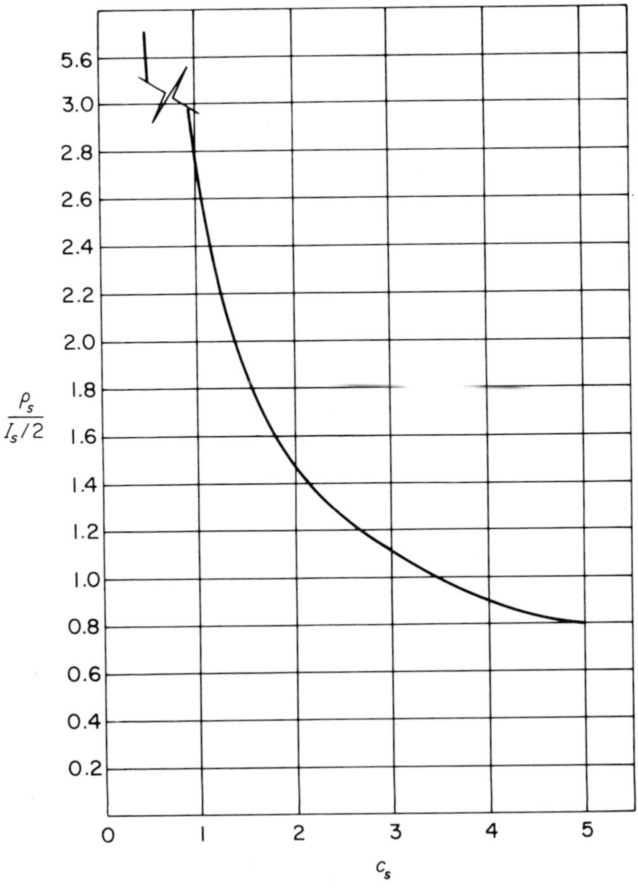

FIG. 8.14 Range resolution ρ_s. Defined as the half-power width of main response of $|q(\tau, o)|$.

system is to have ρ_* (3 db) resolution in slant range and azimuth; all spurious responses in the ambiguity function are to be $-\xi_*$ db (relatively). The RF wavelength is λ_*, and azimuth offset and matched filter processing are assumed.

First, from Fig. 8.9, curves (b) and (d), we find the minimum allowable c_{h*}. The horizontal aperture illumination function is given in Fig. 8.4. From Fig. 8.11, we find $2I_{h*}\rho_*/\lambda_*$ and, hence, determine I_{h*}. Using Eq. (8.5), we find the horizontal aperture length $D_{h*} \equiv 2\Omega_{h*}$.

Second, the PRF f_{S*} is given by Eq. (8.1). $f_{S*} = 4v_* I_{h*}/\lambda_*$.

Third, from Fig. 8.10, e.g., curve (b), we find the minimum allowable c_{v*}. The vertical aperture illumination function is given by Fig. 8.4. Then I_{v*} is found using Eq. (8.2). Using Eq. (8.5), we calculate the vertical aperture length $D_{v*} \equiv 2\Omega_{v*}$.

Fourth, from Fig. 8.13, we find the minimum allowable c_{S*}. The required frequency modulation, assuming a rectangular envelope and a large time-bandwidth product signal modulation, is given in Fig. 8.12. From Fig. 8.14, we find $2\rho_*/I_{S*}$, which determines I_{S*}. By Eq. (8.9), we calculate the RF bandwidth $B_{S*} \equiv 2\Omega_{S*}$.

This completes the specification of the principle SAR system parameters.

It is interesting to note that, if the geometrical situation were such that the inherent suppression of range ambiguities were negligible, the sampling frequency could be found very easily. In any case, the assumption gives a lower bound on f_S. From Figs. 8.8, 8.9, and 8.10, we see that curves (a) of these figures agree closely implying that, for a specified level of spurious response of the ambiguity function, $c_{v*} \approx c_{h*} \equiv \hat{c}$. Note that then the energy of the VPA's in the ambiguous intervals are equal. Using Eqs. (8.1) and (8.2), we find

$$\hat{f}_{S*} = \left(\frac{v v_* \tan \gamma_*}{R_{S*}\lambda_*} \cdot \frac{D_{v*}}{\omega_{h*}}\right)^{1/2},$$

and, in turn,

$$\hat{c} = \frac{\pi}{2}\left(\frac{v \tan \gamma_*}{R_{S*}\lambda_* v_*}\right)^{1/2} D_{v*} D_{h*}.$$

For azimuth offset, $\hat{f}_S$ should be doubled, and $\hat{c}$ halved.

APPENDIX 8A: ALTERNATE CRITERION WITH AZIMUTH OFFSET

In the remarks summarizing the assumptions, it was noted that one might be led to the criterion: *maximize the energy of the one-way voltage pattern in the unambiguous interval, relative to the energy in the adjacent ambiguous intervals.*

In a manner quite analogous to that used to solve the problem posed by the other criterion, we are led to the formulation:

$$\max_{E \in \mathscr{L}^2} \nu$$

where

$$\nu \equiv \frac{\displaystyle\int_{-\frac{1}{2}I_h}^{\frac{1}{2}I_h} \{B_h(u) - [B_h(u + I_h) + B_h(u - I_h)]\}\, du}{\displaystyle\int_{-\infty}^{\infty} B_h(u)\, du}.$$

The integral equation that results is

$$\mu_n f_n = Bf_n,$$

where

$$Bf_n(t) \equiv \int_{-1}^{1} \frac{\sin c(t-s)}{\pi(t-s)} [1 - \cos 2c(t-s)] f_n(s)\, ds.$$

The largest positive characteristic value v_o, the maximum of v, and its associated characteristic function, were found by numerical computation on a digital computer. The resulting aperture functions are shown in Fig. 8.15.

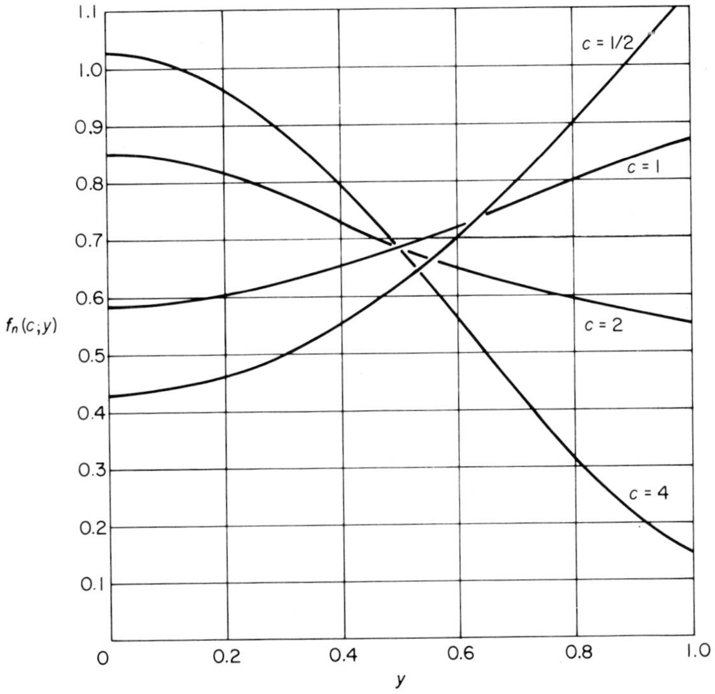

FIG. 8.15 Optimum aperture functions under an alternate criterion.

It is observed that over the range of c commonly used, the solutions are remarkably similar to the solutions of the other maximization problem. When c is small, it is quite clear that the solutions give patterns that have a null in the ambiguous interval.

In the more general problem that takes into account the energy in the even ("compressible") ambiguous intervals, partial results indicate that the solutions are very close to the functions $S_{oo}(c; \cdot)$.

APPENDIX 8B: THE PHASE MODULATION THAT RELATES $|F|$ AND $|\tilde{F}|$

Suppose one has the relation

$$F(t) = F_m(t)e^{i\theta(t)} = \mathscr{F}^{-1}\{\tilde{F}(\omega)\} = \mathscr{F}^{-1}\{\tilde{F}_m(\omega)e^{-i\varphi(\omega)}\}$$

$$= \frac{1}{2\pi}\int_{-\infty}^{\infty} \exp[i\omega t - i\varphi(\omega)]\tilde{F}_m(\omega)\, d\omega.$$

We will be specifically interested in relating $|F(t)| = F_m$ and $|\tilde{F}(\omega)| = \tilde{F}_m(\omega)$. To state precisely when this equation is solvable for specified F_m and $\tilde{F}_m$ appears to be quite difficult. (Certainly the pair must satisfy Parseval's relation, and, for example, they must not have too small a time-bandwidth product.) The method of stationary phase will be used to get an approximate explicit solution; recall that (App. 2A) the essence of this approximate method is that the major contribution of the integrand to the integral occurs at the point(s) where the phase is stationary. For want of a name, the class of functions amenable to the approximations required may be said to be of "large time-bandwidth product" (in part by analogy to the linear FM radar signal). Having arrived at F_m and $\tilde{F}_m$, this product (using, e.g., variance as a measure of spread), and hence the appropriateness of the method, can be estimated. In any case, having the signal, the realized ambiguity function and its cross sections can be computed and compared with the proposed cross sections.

The stationary point(s) is determined by

$$\frac{d}{d\omega}\{\omega t - \varphi(\omega)\} = 0. \tag{8B.1}$$

Thus, the stationary point(s) $\omega = \lambda(t)$ is such that $t = +\varphi'(\lambda)$. It is assumed that for each t, λ is unique (that is, $\lambda(t)$ is a single-valued function). One expands the phase in a Taylor series about λ and makes the following assumptions: the power series will be truncated at the second-order term, and it is assumed that in the resulting ω interval, where this is a good approximation to the phase, $\tilde{F}_m(\omega)$ is approximately a constant, and the resulting Fresnel integral "picks up" most of its value.

Thus, in detail, we take

$$\omega t - \varphi(\omega) \approx \lambda t - \varphi(\lambda) - \tfrac{1}{2}\varphi''(\lambda)(\lambda - \omega)^2, \qquad \lambda - \varepsilon \le \omega \le \lambda + \varepsilon,$$

where ε is such that the polynomial approximation is "good." Then,

$$F(t) \approx \frac{1}{2\pi}\exp[i\lambda t - i\varphi(\lambda)]\int_{s-\varepsilon}^{\lambda+\varepsilon} \exp[-\tfrac{1}{2}i\varphi''(\lambda)(\lambda - \omega)^2]\tilde{F}_m(\omega)\, d\omega.$$

If $\tilde{F}_m(\omega) \approx F_m(\lambda)$, $\omega \varepsilon (\lambda - \varepsilon, \lambda + \varepsilon)$, then,

$$F(t) \approx \frac{1}{2\pi} \exp[i\lambda t - i\varphi(\lambda)] \tilde{F}_m(\lambda) \int_{\lambda-\varepsilon}^{\lambda+\varepsilon} \exp[-i\tfrac{1}{2}\varphi''(\omega)(\lambda - \omega)^2] \, d\omega.$$

And, finally, if the integral

$$\int_{-\infty}^{\infty} \exp[-i\varphi''(\lambda)(\omega - \lambda)^2] \, d\omega \approx \int_{\lambda-\varepsilon}^{\lambda+\varepsilon} \exp[-\tfrac{1}{2}i\varphi''(\lambda)(\omega - \lambda)^2] \, d\omega,$$

which requires that $\varphi''(\lambda)$ be "large," one has

$$F(t) \approx \frac{1}{(2\pi)^{1/2}} \frac{\tilde{F}_m(\lambda)}{|\varphi''(\lambda)|^{1/2}} \exp[i\lambda t - i\varphi(\lambda) - i(\pi/4)].$$

One observes

$$\theta'(t) \approx \lambda' - \varphi'(\lambda) \cdot \lambda' = \lambda(t)$$

where [using Eq. (8B.1)] $\lambda(t)$ is the frequency modulation of the signal $F(t)$. Further,

$$F_m(t) \approx \frac{1}{(2\pi)^{1/2}} \frac{\tilde{F}_m(\lambda)}{|\varphi''(\lambda)|^{1/2}}.$$

Suppose one looks for solution for this equation among φ such that φ'' is nonnegative; it turns out that such a solution exists under reasonable conditions. Then one is lead to consider the differential equation for

$$F_m(t) = \frac{1}{(2\pi)^{1/2}} \frac{\tilde{F}_m[\lambda(t)]}{(\varphi''[\lambda(t)])^{1/2}}, \tag{8B.2}$$

for t, such that $F(t) \neq 0$. Using Eq. (8B.1), Eq. (8B.2) becomes

$$F_m^{\,2}[\varphi'(\lambda)]\varphi''(\lambda) = \frac{1}{2\pi} F_m^{\,2}(\lambda).$$

Setting $g = F_m^{\,2}$, $g_* = \tilde{F}_m^{\,2}$, one has

$$g[\varphi'(\lambda)]\varphi''(\lambda) = \frac{1}{2\pi} g_*(\lambda),'$$

or

$$\frac{d}{d\lambda} G[\varphi'(\lambda)] = \frac{1}{2\pi} g_*(\lambda),$$

where

$$G(\zeta) = \int^{\zeta} g(\xi) \, d\xi.$$

Integrating, one has

$$G[\varphi'(\lambda)] = \frac{1}{2\pi}\int^{\lambda} g_*(\xi)\,d\xi + c \equiv \frac{1}{2\pi}\,G_*(\lambda) + c, \tag{8B.3}$$

where c is an arbitrary constant. Again, using Eq. (8B.1), Eq. (8B.2) becomes

$$G(t) = \frac{1}{2\pi}\,G_*[\lambda(t)] + c.$$

Proceeding formally, one then has

$$\lambda(t) = G_*^{-1}[2\pi G(t) - c].$$

This equation is to be satisfied for all t such that $\tilde{F}_m[\lambda(t)] \neq 0$. The range of G_*—which is the domain of G_*^{-1}— is $(0, 2\pi)$, and the range of G is $(0, 1)$; so one takes $c = 0$. G_* is certainly monotone nondecreasing, and if, in fact, it is strictly increasing, then G_*^{-1} exists as a single-valued function. Thus, one has for the frequency modulation of the complex signal modulation

$$\lambda(t) = G_*^{-1}\{2\pi G(t)\}. \tag{8B.4}$$

APPENDIX 8C: SOLUTION OF NONLINEAR DESIGN PROBLEM FOR MATCHED FILTERING

In Sec. 8.2, in discussing design criteria based on the SAR ambiguity function, the design criterion for the azimuth pattern settled upon had its greatest appeal when the phase—but not the amplitude—was matched. We now want to show that, generally speaking, the reasons for extending this criterion to matched (phase and amplitude) processing were correct. When matched filtering is employed, the following criterion is reasonable:

> The design criterion for the one-way voltage pattern in azimuth is to maximize the squared energy in the unambiguous interval.

The mathematical problem is now readily seen to be

$$\max_{f \in \mathscr{L}^2} (Tf, f)_1, \qquad \text{subject to } \|f\|_1{}^2 = \text{constant}, \tag{8C.1}$$

instead of Eq. (8.5), where the nonlinear operator T is defined by

$$Tf(\omega) \equiv \int_{-1}^{+1} dx \int_{-1}^{+1} dy \int_{-1}^{+1} dz\,\frac{\sin C(\omega - x + y - z)}{\pi(-x + y - z)}\,f(x)\bar{f}(y)f(z), -1 \le \omega \le 1.$$

Again the desired solution is the characteristic function(s) associated with the largest characteristic value of the integral equation

$$Tf = \lambda f \tag{8C.2}$$

[compare Eq. (8.7)]. Equation (8C.2) was solved numerically, and the maximum of $(Tf, f)_1$ is shown in Table 8.1, with the solutions plotted in Fig. 8.16.

We see that the solutions would be practically indistinguishable for $C \geq 3$, and $(Tf, f)_1$ changes only by a few parts in 10^{-4} for $C \geq 3$, a practically negligible amount. This is a somewhat surprising result.

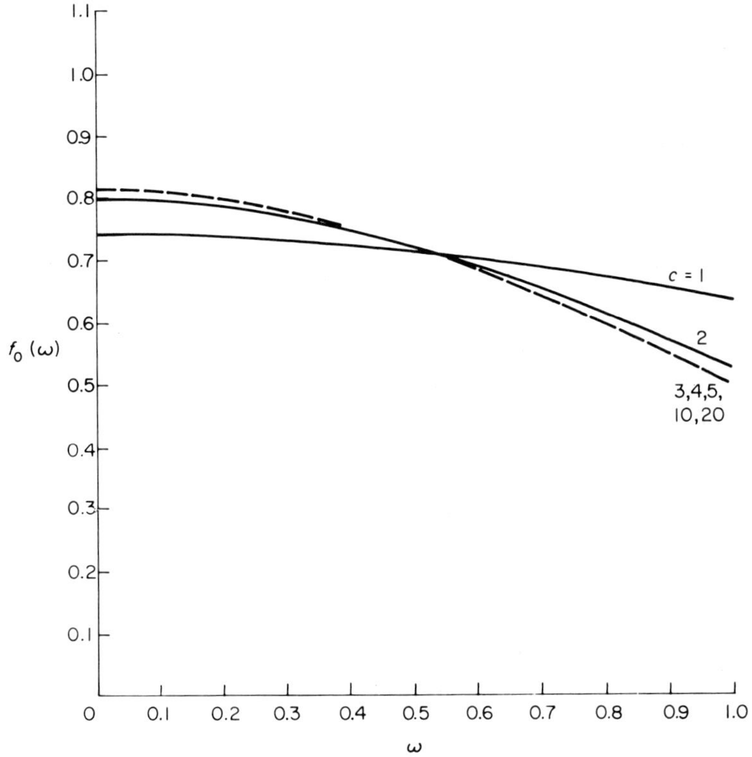

FIG. 8.16 The functions, f, that maximize $(Tf, f)_1$.

TABLE 8.1

THE MAXIMUM OF $(Tf, f)_1$

c	$(Tf, f)_{max}$	c	$(Tf, f)_{max}$
0.5	0.6029884	3.5	1.3733643
1.0	1.0393267	4.0	1.3733809
1.5	1.2674347	4.5	1.3735621
2.0	1.3515171	5.0	1.3738848
2.5	1.3711825	10.0	1.3742212
3.0	1.3733223	20.0	1.3742697

As a consequence, we conclude that we do not really care for this criterion. For, since all these illuminations—and hence their far-field patterns over the unambiguous azimuth angular interval—are so nearly constant, the relative height of the azimuth ambiguities will be closely proportional to the *energy* of the pattern in the ambiguous intervals. But we know from the results of the present chapter that, by proper choice of the illumination functions, this energy can be markedly reduced, as C increases above 3.

COMMENTS AND REFERENCES

The discussion of the present chapter closely follows Harger (1965b). The prolate spheroidal wave functions were thoroughly discussed in an admirable series of papers initiated by Slepian and Pollock (1961). The classical linear variational problem is discussed by, e.g., Reisz and Sz.-Nagy (1955). The nonlinear variational problem was discussed by Harger and Crimmins (1968).

The problem of finding a phase modulation that relates the specified moduli of a (proposed) Fourier transform pair has been approximately solved under the large TBP condition by Harger (1963), and Fowle (1964).

IX

The Ambiguity Function

9.1. INTRODUCTION

As was observed at the very outset, one of the apparently fundamental restraints on SAR systems is the upper bound on the unambiguous range interval for fixed vehicle velocity and azimuth resolution [Eq. (1.1)]. We wish now to investigate this apparent limitation. The problem of how to combat this limitation is called "the SAR ambiguity problem"; we shall define a more precise statement of it below.

The ambiguity problem is the most difficult and subtle problem of the SAR system. Paradoxically—but not unusually—the problem is in an unsatisfactory theoretical state, and yet satisfactory practical "solutions" are known.

The practical solutions seem to depend on the fact that, generally speaking, a radar image can be "dispersed"—by, e.g., linear dispersion operations, as we have discussed above—and added to an image with essentially no observable degradation to the latter image. In fact, several such dispersed images can be added to an image with little visible degradation of the latter image. For example, one can disperse an image by free space diffraction: if the dispersion factor is large, the dispersed image will be seen to be very smooth relative to its original fluctuations, so much so that it can be considered merely a "gray level" or uniform background. This explains its negligible effect when added to an image.

This experimental observation implies that the nature of the envelope of the image is greatly different from the nature of the envelope of the dispersed image. Since the dispersion operation can be viewed as the output of a LIV operation whose transfer function is purely a quadratic phase (see Chapters III and V) when the input is the image, *the spectral densities of the image and the dispersed image are the same*. This implies that the image is not a Gaussian process: for if the image were a Gaussian process, then the dispersed image would also be, and, since the spectral density completely determines a (zero mean) Gaussian random process, they would be indistinguishable random processes.

In such a case, the ordinary least squares criterion is not very useful. For, consider the situation shown in Fig. 9.1: this is the block diagram of the

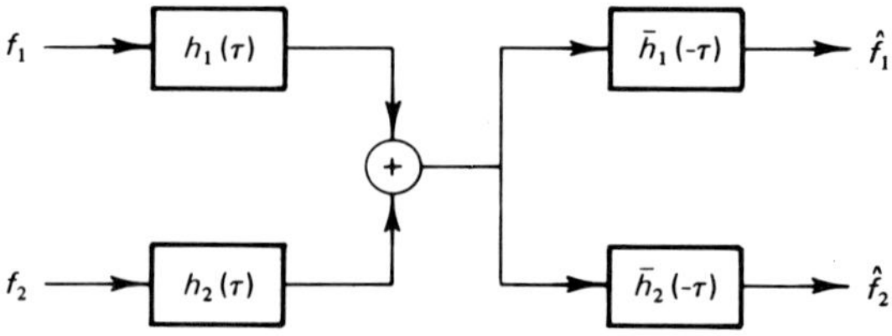

FIG. 9.1 An ambiguity suppression scheme suitable for imaging twice the unambiguous range interval.

range channel of an SAR system that can satisfactorily image twice the unambiguous range (as we will discuss in Sec. 9.3.2). The LIV filters h_1 and h_2 are dispersion filters, $\bar{h}_1$ and $\bar{h}_2$ are their matched/compression filters, f_1 can represent the terrain in the near unambiguous range interval, and f_2 can represent the terrain in the far unambiguous range interval. They are received simultaneously additively, and we know experimentally that, e.g., $\hat{f}_1$ can be a very good image with f_2 present, hardly distinguishable from the image with f_2 absent. Yet, if we were to use MSE as a criterion of image quality, we would receive a much different impression. For suppose $|\tilde{\bar{h}}_i|^2\, \tilde{f}_i = \tilde{f}_i$, for $i = 1, 2$—i.e., the "top" and "bottom" channels of Fig. 9.1, transmit f_1 and f_2, respectively, perfectly—and $|\tilde{h}_1|^2\, |\tilde{h}_2|^2\, |\tilde{f}_i|^2 = |\tilde{f}_i|^2$, for $i = 1, 2$. (These assumptions leave the phases of $\tilde{h}_1$ and $\tilde{h}_2$ free, and that is all we need to obtain the experimental effect we are discussing.) Then, e.g.,

$$\tilde{\hat{f}}_i = \tilde{f}_i + \tilde{\bar{h}}_i \tilde{h}_j \tilde{f}_j,$$

and the error $\varepsilon_i \equiv f_i - \hat{f}_i$ has spectrum $\tilde{\bar{h}}_i \tilde{h}_j \tilde{f}_j$, so that the mean square error, MSE_i, is

$$\mathrm{MSE}_i \equiv \int |\varepsilon|^2 = \int |f_j|^2, \qquad (i, j) = (1, 2), (2, 1).$$

Since normally we would have $\int |f_1|^2 = \int |f_2|^2$, our result states that the MSE in estimating f_i is the same as it would be if we took the zero function as our estimate of f_i, for $i = 1, 2$!

In order to find a criterion sufficiently simple to allow fruitful discussion, we shall use the ML criterion for a point object received in additive noise—

as introduced in Chapter IV—and use the ambiguity function already discussed (in simplified form) in Chapter VIII. We must, of course, orient the discussion properly to account for the fact that we are mapping a strip of distributed reflectivity. Therefore, we once again consider the design of the impulse response of the SAR system but now attempt to generalize the discussion somewhat.

The fact that we are mapping a strip of distributed reflectivity, with special interest in a specific spatial bandwidth of the distributed reflectivity, seems to require that the signal modulation be periodic or "nearly" periodic. Certainly we must emit at least one pulse during each synthetic aperture distance—else there would be very little signal to process at all. We have already given (Chapter II) the heuristic reason for sampling on the order of the spatial bandwidth.

One can impose a periodic requirement on the signal modulation in a more direct, although a little contrived, way. As we saw in Chapter IV, the essential structure of the ML processor is given by $|\int Z(t)\bar{S}(t)\,dt|^2$, where (as we will show immediately below) in the SAR system $S(t) = F(t-\tau)A(t-\zeta)$, the range and azimuth positions being represented by τ and ζ, respectively. Now suppose that

$$Z(t) = \sum b_n F(t-\tau_o)A(t-n\rho),$$

the signal scattered by a row of point scatterers, all at range delay τ_o and spaced ρ apart in azimuth. A purpose of the radar is, of course, to estimate the $\{b_n\}$—which for simplicity we take real and equal —and we require that the SAR system do "equally well" for each b_n. As we shall discuss in Sec. 9.2, a good measure—from both an accuracy and a resolution point of view—is

$$-\partial^2 \left| \int Z(t)\bar{S}(t)\,dt^2 \right| / \partial b_n{}^2, \qquad \text{for } (\tau, \zeta) = (\tau_0, n\rho).$$

We find this derivative to be

$$\left| \int F(t-\tau_0)\bar{F}(t-\tau)A(t-n\rho)\bar{A}(t-\zeta)\,dt \right|^2,$$

and we want them to be equal for all n when $(\tau, \zeta) = (\tau_o, n\rho)$. Possible solutions are:

(1) $|A(t)|^2 \equiv$ constant,

(2) $|F(t)|^2 \equiv$ constant,

(3) $|A(t)|^2$ of period ρ,

(4) $|F(t)|^2$ of period ρ.

The first two solutions are obviously unacceptable, the third certainly not feasible, and thus we are led to requiring that the magnitude $|F|^2$ be of period ρ. The parameter ρ is, of course, a resolution parameter.

9.2. GENERAL PROPERTIES

Definition. The form of the received temporal signal scattered by a point object was found in Sec. 3.2 [Eq. (3.9)] to be, in complex form,

$$S(t) = g_o\left(\frac{ik}{4\pi R_m}\right)^2 F(t - \tau_m - \tau_o)B_v(\zeta_o)B_h(x_o - vt; \zeta_o), \tag{9.1}$$

assuming the reflectivity density of the scatterer is g_o and its position on the terrain plane is (ζ_o, x_o). (See Fig. 3.1 for the usual geometry.) We have defined the delay $2\zeta_o \cos \gamma_o/v = 2y_o/v$ to be τ_o. It will be convenient to simplify the the notation at this point: we set

(1) $\mu \equiv x/v$,

(2) $A(t; \tau_o) \equiv B_h(vt; \zeta_o)$,

(3) $a(\tau) \equiv |B_v(\tau)|^2/|B_v(0)|^2$,

(4) $b_o \equiv g_o(ik/4\pi R_m)^2$.

Now

$$S(t) = b_o B_v(\zeta_o)F(t - \tau_m - \tau_o)A(t - \mu_o; \tau_o). \tag{9.1a}$$

Recalling the ML processing procedure (Sec. 4.3.3), we form the inner product of the data and the signal—Eq. (9.1a)—with "test" parameters $(\tilde{\tau}, \tilde{\mu})$, take the modulus squared, and divide by the squared norm of "test" signal—Eq. (4.10). We assume that the processor is set up with test parameters $(\tilde{\tau}, \tilde{\mu}) = (0, 0)$, and we shall investigate the behavior of the processor as the location (τ, μ) of the point scatterer is varied. The function is

$$|q(\tau, u)|^2 = |a(\tau)|^2 \frac{\left|\int_T F(t - \tau)\bar{F}(t)A(t - \mu; \tau)\bar{A}(t; 0)\, dt\right|^2}{\left(\int_T |F(t)A(t; 0)|^2\, dt\right)^2}. \tag{9.2}$$

Definition 9.1. The function $|q|^2$ defined by Eq. (9.1) is the *ambiguity function* of the SAR system.

We have ignored the uninteresting fixed delay $\tau_m = 2R_m/v$ and squared the normalization factor in Eq. (9.2) to conveniently scale $|q|^2$—see Prop. 9.1.

We assume that $|a(\tau)^2 \leq |a(0)|^2$: this is rather inessential but avoids useless complication about the obvious variation of the vertical gain pattern. We shall often write $A(t; \tau)$ simply as $A(t)$, suppressing its mild dependence on τ.

Bounds. The first properties of $|q|^2$ that we find are bounds.

Property 9.1. "$|q(\tau, \mu)|^2 \leq |q(0, 0)|^2 = 1.$"

Proof. We apply the Schwarz inequality

$$\left|\int f\bar{g}\right| \leq \int |f|^2 \cdot \int |g|^2$$

to the numerator of the right side of Eq. (9.2) with

$$f(t) \equiv F(t - \tau)A(t - \mu)$$

and

$$g(t) \equiv F(t)A(t). \qquad \text{Q.E.D.}$$

We recall that equality is achieved in (the Schwarz inequality) if and only if $f \equiv \alpha g$, for α *a* complex constant. That is, $|q|^2$ achieves the value unity, if and only if $F(t - \tau)A(t - \mu) = F(t)A(t)$: this is obviously true—as indeed we want—when $(\tau, \mu) = 0, 0)$. If it is true for another point (τ_1, μ_1), then (τ_1, μ_1) is said to be "completely ambiguous" with (0, 0).

In the statement of the next property we denote the denominator of Eq. (9.2) by D.

Property 9.2.

$$\text{“}\,|q(\tau, \mu)|^2 \leq \frac{|a(\tau)|^2}{D} (F_m{}^2 * F_m{}^2)(\tau) \cdot |A|^2 * |A|^2(\zeta).\text{”}$$

Proof. Just as in proving Prop. 9.1, we use the Schwarz inequality, except now we set

$$f(t) \equiv F(t - \tau)\bar{F}(t)$$

and

$$g(t) \equiv A(t - \zeta)\bar{A}(t) \qquad (\text{recall } F_m \equiv |F|). \qquad \text{Q.E.D.}$$

This is a rather simple form for a bound, easily interpreted because it is simply a product of convolutions. It is not as "tight" as possible for large τ

because the phase modulation of A depends on τ, also causing a lower level than that given, as we have often mentioned above.

Property 9.2 gives us directly an indication of "clear" regions—regions where $|q|^2$ is identically zero.

Corollary 9.2.1. "If F_m is a periodic repetition, with period T_s of function F_{om} whose support lies in $(-\frac{1}{2}T_o, \frac{1}{2}T_o)$, where $2T_o < T_s$, then there are μ-strips in the (τ, μ)-plane, centered on $\tau = (n + \frac{1}{2})T_s$, for $n = 0, \pm 1, \ldots$, and of τ-width $T_s - 2T_o$, where $|q(\tau, \mu)|^2$ is identically zero."

Proof: Since the support of the convolution $F_{om}^2 * F_{om}^2$ lies in $(-T_o, T_o)$, the result follows trivially from Prop. 9.2. Q.E.D.†

We note that dispersed pulse modulations reduce the cleared area relative to nondispersed pulse modulations. In the same way, the SAR system ambiguity function utilizes a $|B_h|^2$ of greater extent than the $|B_h|^2$ of a "brute force" sidelooking radar, so that the bound $A_h{}^2 * A_h{}^2$ is of greater significant extent. We also remark that Cor. 9.2.1 assumes only that F_m—not necessarily F—has the hypothesized periodicity.

Periodic properties. As remarked above, periodic or nearly periodic signal modulations appear to be mandatory in SAR systems. The properties of the ambiguity function for periodic modulations is therefore of special interest.

Property 9.3. "If F is periodic with period T_s, and the τ-dependence of A ignored, then $|q(\tau, \mu)|^2/|a(\tau)|^2$ is periodic in τ with period T_s."

Proof. This is trivial, from the definition of $|q|^2$ by Eq. (9.2). Q.E.D.

The periodicity in range delay τ of $|q|^2/|a(\tau)|^2$ due to the periodicity of the signal modulation F, may be said to characterize the "range ambiguity problem." For, if we attempt to measure delays greater than T_s, ambiguity arises—that is, an object at delay T_s gives a response that is ambiguous with the response of an object at delay 0, e.g.; since their respective scattering amplitudes are unknown, the different gains, $|a(T_s)|^2$ and $|a(0)|^2$, assigned them are not useful in distinguishing them. It is true, as we have often mentioned above, that the azimuth modulation A depends on τ: but this does not provide sufficient information to practically distinguish, e.g., a point scatterer at delay T_s from a "distributed" object at delay 0.

† Strictly, $F(t-\tau)F(t)$ must be integrable: it is sufficient to truncate F_m to any finite interval.

We shall find the following definitions useful:

Definition 9.2.

$$\tilde{G}_{T_s}(\omega; \tau) \equiv \int_{T_s} e^{-iwt} F(t-\tau)\bar{F}(t)\, dt, \tag{9.3}$$

and

$$\tilde{H}(\omega; \mu) \equiv \int e^{-iwt} A(t-\mu)\bar{A}(t)\, dt. \tag{9.4}$$

Lemma 9.4.1. "If $\tilde{G}_{Ts}$ and $\tilde{H}$ are defined by Eqs. (9.3) and (9.4), respectively, then

$$\tilde{G}_{T_s}(\omega; \tau) = \frac{1}{2\pi} \int e^{-iu\tau} \bar{\tilde{F}}(u-\omega) \tilde{F}_{T_s}(u)\, du, \tag{9.3a}$$

where

$$\tilde{F}_{T_s}(u) \equiv \int_{T_s} e^{-iut} F(t)\, dt,$$

and

$$\tilde{H}(\omega; \mu) = \frac{1}{2\pi} \int e^{-i\mu r} \bar{\tilde{A}}(r-\omega) \tilde{A}(r)\, dr." \tag{9.4a}$$

We can now state and prove:

Property 9.4. "If F has period T_s and $\omega_s \equiv 2\pi/T_s$, then

$$|q(\tau, \mu)|^2 = |a(\tau)|^2 \frac{|\sum \tilde{G}_{T_s}(n\omega_s; \tau)\tilde{H}(n\omega_s; \mu)|^2}{(\sum \tilde{G}_{T_s}(n\omega_s; 0)\tilde{H}(n\omega_s; 0))^2}. \tag{9.5}$$

If additionally, $\tilde{A}(\omega) \equiv 0, \quad |\omega| > \frac{1}{2}\omega_s$, then

$$|q(\tau, \mu)|^2 = |a(\tau)|^2 \frac{\left|\int_{T_s} F(t-\tau)\bar{F}(t)\, dt\right|^2 \cdot \left|\int A(t-\mu)\bar{A}(t)\, dt\right|^2}{\left(\int_{T_s} |F(t)|^2\, dt\right)^2 \cdot \left(\int |A(t)|^2\, dt\right)^2}." \tag{9.6}$$

Proof. We define $G(t; \tau) \equiv F(t-\tau)\bar{F}(t)$ and note that the periodicity T_s of F implies the same for G in t (and τ also). Therefore, G as a function of t has a Fourier series expansion:

$$G(t; \tau) = \frac{1}{T_s} \sum \tilde{G}_{T_s}(n\omega_s; \tau) \exp in\omega_s t, \tag{9.7}$$

where $\omega_s \equiv 2\pi/T_s$, and $\tilde{G}_{Ts}$ denotes the Fourier transform of $G(t)$ defined nonzero only on $(-\frac{1}{2}T_s, \frac{1}{2}T_s)$—see Eq. (9.3) of Def. 9.2. Substituting Eq. (9.7) into $|q|^2$, we get the form Eq. (9.5) if $\tilde{H}(\omega; \mu)$ is defined as in Eq. (9.4) of Def. 9.2.

Finally, from Eq. (9.4a) of Lemma 9.4.1, we see that if $\tilde{A}(\omega) \equiv 0$, for $|\omega| > \frac{1}{2}\omega_s^2$, then $\tilde{A}(r - n\omega_s)\tilde{A}(r) \equiv 0$, for $n \neq 0$. Hence Eq. (9.5) reduces in this case to Eq. (9.6). Q.E.D.

The result, Eq. (9.6), given by Prop. 9.4 is a very simple form: it occurs when the azimuth modulation is adequately sampled—that is, when the second hypothesis is satisfied. This, of course, is never strictly, but only ideally, true in the SAR system. We list the attributes of Eq. (9.6):

1. The range resolution properties are determined, except for the second-order dependence of A on τ, by the signal modulation F only;
2. The azimuth resolution properties are determined by the azimuth modulation A only.

Threshold accuracy. In many radar systems it is reasonable to assume that there is one "point" scattering object, and that its parameters are to be estimated in the presence of weak additive noise—usually Gaussian and/or white. The maximum likelihood procedure leads to well-known bounds on the variance of the error of the estimates of the unknown parameters—the so-called Cramer-Rao lower bound. The bound involves the mixed second-order partial derivatives of the likelihood ratio, or functional, with respect to the "trial" parameters. This investigation for weak additive noise has been termed the "threshold accuracy" problem.

Such considerations are not of major concern in systems such as the SAR system, for several reasons: (1) absolute accuracy is generally not of interest, but rather relative accuracy or distortion is important; (2) the single scatterer assumption is of course not true; (3) undue attention to the likelihood functional—that is, the impulse response of the processor optimum under the maximum likelihood criterion—at the origin of parameter space can lead to a wild distortion of the design in the remainder of the parameter space.

It remains true that the second-order partial derivatives of the likelihood functional at $(\tau, \mu) = (0, 0)$ give a partial measure of the resolution properties of the system and are therefore of interest. For example $-\partial^2|q|^2/\partial\tau^2$, at $(0, 0)$ is a measure of the "sharpness" of the impulse response in the delay direction; similarly for $-\partial^2|q|^2/\partial\mu^2$. The derivative $-\partial^2|q|^2/\partial\tau\,\partial\mu$ is a measure of the sharpness of $|q|^2$ in an off-axis direction.

Property 9.5. "Neglecting the τ-variation of a and A, a measure of resolution is given by

$$\frac{\partial^2}{\partial \tau^2}|q|^2\bigg|_{(0,0)} = \frac{2}{D^2}\left[D \operatorname{Re}\int F''(t)\bar{F}(t)\,|A(t)|^2\,dt + \left|\int F'(t)\bar{F}(t)\,|A(t)|^2\,dt\right|^2\right],$$

$$\frac{\partial^2}{\partial \mu^2}|q|^2\bigg|_{(0,0)} = \frac{2}{D^2}\left[D \operatorname{Re}\int |F(t)|^2\,A''(t)\bar{A}(t)\,dt + \left|\int |F(t)|^2\,A'(t)\bar{A}(t)\,dt\right|^2\right],$$

$$\frac{\partial^2}{\partial \tau\,\partial \mu}|q|^2\bigg|_{(0,0)} = \frac{2}{D^2}\operatorname{Re}\left[D\int F'(t)\bar{F}(t)A'(t)\bar{A}(t)\,dt\right.$$

$$\left. + \int F'(t)\bar{F}(t)\,|A(t)|^2\,dt \cdot \int |F(t)|^2\,A'(t)A(t)\,dt\right],$$

where

$$D \equiv \int |F(t)|^2\,|A(t)|^2\,dt."$$

Proof. We simply carry out the indication differentiations. Q.E.D.

We have already remarked that the delay variation of $|a|^2$ and A is of no use as a delay resolution property, and they certainly have negligible effect on $\partial^2|q|^2/\partial\tau^2$.

Corollary 9.5.1. "If F and A are even functions, both about any point $t = t_*$, then the partial derivatives of Prop. 9.5 become

$$\frac{\partial^2}{\partial \tau^2}|q|^2\bigg|_{(0,0)} = \frac{2}{D}\operatorname{Re}\int F''(t)\bar{F}(t)|A(t)|^2\,dt,$$

$$\frac{\partial^2}{\partial \mu^2}|q|^2\bigg|_{(0,0)} = \frac{2}{D}\operatorname{Re}\int |F(t)|^2\,A''(t)\bar{A}(t)\,dt,$$

$$\frac{\partial^2}{\partial \tau\,\partial \mu}|q|^2\bigg|_{(0,0)} = \frac{2}{D}\operatorname{Re}\int F'(t)\bar{F}(t)A'(t)\bar{A}(t)\,dt,$$

where

$$D \equiv \int |F(t)|^2\,|A(t)|^2\,dt."$$

Proof. If a complex function is even, then its derivative is, as for a real function, an odd function, and its magnitude is also even. Hence Cor. 9.5.1. follows trivially from Prop. 9.5. Q.E.D.

In the following corollary, we ignore the τ-dependence of a and A.

Corollary 9.5.2. "If F has periodicity T_s, and $\tilde{A}(\omega) \equiv 0$, for $|\omega| > \frac{1}{2}\omega_s \equiv \pi/T_s$, then the partial derivatives of Prop. 9.5 may be written:

$$\frac{\partial^2}{\partial\tau^2}|q|^2\bigg|_{(0,0)} = \frac{2}{D_F{}^2}\left[D_F \operatorname{Re}\int_{T_s} F''(t)\bar{F}(t)\,dt + \left|\int_{T_s} F'(t)\bar{F}(t)\,dt\right|^2\right],$$

$$\frac{\partial^2}{\partial\mu^2}|q|^2\bigg|_{(0,0)} = \frac{2}{D_A{}^2}\left[D_A \operatorname{Re}\int A''(t)A(t)\,dt + \left|\int A'(t)\bar{A}(t)\,dt\right|^2\right],$$

$$\frac{\partial^2}{\partial\tau\,\partial\mu}|q|^2\bigg| \equiv 0,$$

where

$$D_A \equiv \int |A(t)|^2\,dt,$$

and

$$D_F \equiv \int_{T_s} |F(t)|^2\,dt."$$

Proof. We first establish the form for $\partial^2|q|^2/\partial\tau^2$ at (0, 0). Since μ is not involved, we set $\mu \equiv 0$ at the outset, in Eq. (9.6) of Prop. 9.4:

$$|q(\tau, 0)|^2 \equiv \frac{1}{D_F{}^2}|q_F(\tau)|^2 = \frac{1}{D_F{}^2}\left|\int_{T_s} F(t-\tau)\bar{F}(t)\,dt\right|^2.$$

Therefore,

$$\frac{\partial^2}{\partial\tau^2}|q(\tau,0)|^2 = \frac{1}{D_F{}^2}\frac{\partial^2}{\partial\tau^2}|q_F(\tau)|^2 = \frac{2}{D_F{}^2}[\operatorname{Re}(q_{F_{\tau\tau}}\bar{q}_F) + |q_{F_\tau}|^2]. \quad (9.8)$$

We compute:

$$q_F(0) = D_F,$$

$$q_{F_\tau}(0) = -\int_{T_s} F'(t)\bar{F}(t)\,dt,$$

$$q_{F_{\tau\tau}}(0) = +\int_{T_s} F''(t)\bar{F}(t)\,dt.$$

Hence, substituting these calculations into Eq. (9.8) gives the desired form for $\partial^2|q|^2/\partial\tau^2$ at (0, 0).

The proof of the form for $\partial^2|q|^2/\partial\mu^2$ at (0, 0) proceeds analogously, and the form is the analog of $\partial^2|q|^2/\partial\tau^2$.

The fact that the mixed partial derivative $\partial^2|q|/\partial\tau\partial\mu$ is zero at (0, 0), follows directly from the factored form of Eq. (9.6) of Prop. 9.4:

$$\frac{\partial}{\partial\tau\,\partial\mu}|q|^2 \propto \frac{\partial}{\partial\tau}|q_F(\tau)|^2\,\frac{\partial}{\partial\mu}|q_A(\zeta)|^2.$$

For both $|q_F|^2$ and $|q_A|^2$ have a maximum at zero, so their derivatives there are necessarily zero. Q.E.D.

9.3. THE RANGE EXTENSION PROBLEM

There are many problems that one could pose, centered on the Def. 9.1 of the ambiguity function, as defined by Eq. (9.2), and they all can be loosely termed "ambiguity problems of the SAR system." Since there are only, at most, computational difficulties involved in finding $|q|^2$, given F and A, a more interesting class of problems is the "synthesis" problems: find F and A to produce, either precisely or approximately, a desired form $|p|^2$. Synthesis problems are, therefore, concerned at the outset with the "realizability problem": given a function $|p|^2$, (1) is it an ambiguity function, and (2) if not, can it be approximated sufficiently well by an ambiguity function? Given the answers to these questions—which are not known, by the way—there remains the problem of finding F and A from $|q|^2$; the freedom of the phase of q would enlarge the class of solutions F and A, but there are no known methods of doing this.

Even if there existed answers to these questions, there remains the problem of characterizing the answers and proposed functions $|p|^2$ in a sufficiently simple way. A casual glance at the ambiguity functions discussed in Chapter VIII (Figs. 8.6 and 8.7), indicates the complexity of specifying useful ambiguity functions.

We shall have to be content here with a simpler "ambiguity problem" but one still of great practical interest. The discussion of Chapter VIII demonstrated that, if we are content to remain within the ambiguity bound set by Eq. (1.1), then the ambiguity problem "control the level of all spurious responses for prescribed resolution" was satisfactorily solved. There remains, therefore, the logical ambiguity problem of whether or not it is possible to extend the range interval that is imaged. We shall state this problem in the following way:

> Suppose that one has periodic signal modulation that produces the desired ambiguity function $|q_D|^2$. Is it possible to change the signal modulation in alternate periods so as to remove the first range ambiguity, without introducing azimuth ambiguities, and while otherwise preserving $|q_D|^2$?

The final phrase of the statement of the problem will become clear when we take it up in Sec. 9.3.1: we shall find that if we are, as in practice, restricted to using pulse modulations that are very short relative to the sampling interval, then the answer is "no"!

However, we shall see in Sec. 9.3.2, that the answer to this problem is a practical "yes," if we are willing to accept range and azimuth ambiguities of a very low level.

9.3.1. The Ambiguity Avoidance Problem

Suppose that we have a signal modulation of period T_S that produces a satisfactory ambiguity function $|q_D|^2$ except for the presence of the expected range ambiguity response at $\tau = \pm T_S$. That is, we are satisfied with $|q_D|^2$ for $|\tau| < \frac{1}{2}T_S$, but not for $\frac{1}{2}T_S < \tau < \frac{3}{2}T_S$. We propose to change the signal modulation on alternate pulses to see if this range ambiguity can be removed.

The periodicity of the signal modulation is now $2T_S$; from Eq. (9.5) of Prop. 9.4— and neglecting constants—we have

$$|q(\tau, \mu)|^2 = |\sum \tilde{G}_{2T_S}(\tfrac{1}{2}n\omega_s\, ; \tau)\bar{\tilde{H}}(\tfrac{1}{2}n\omega_s\, ; \mu)|^2.$$

Since the original ambiguity function $|q_D|^2$ was acceptable, we assume that $\tilde{A}(\omega) \equiv 0$, for $|\omega| > \frac{1}{2}\omega_S$, and hence that $\tilde{H}(\frac{1}{2}n/\omega_S; \mu) \equiv 0$, for $|n| > 1$—refer to Eq. (9.4a). Also, for the same reason, we may assume that the $\tilde{H}(\frac{1}{2}n/\omega_S; \mu)$ have, to a very good approximation, disjoint support in μ. With these assumptions, we may write

$$\left|q(\tau, \mu)\right|^2 = \sum_{n=-1}^{+1} |\tilde{G}_{2T_s}(\tfrac{1}{2}n\omega_s\, ; \tau)|^2 \cdot |\tilde{H}(\tfrac{1}{2}n\omega_s\, ; \mu)\Big|^2. \tag{9.9}$$

Since F has period $2T_S$, so does $|q|^2$, and we only need examine $|q|^2$ and, hence, $|G_{2T_S}|^2$, for $-\frac{1}{2}T_S \leq \tau \leq \frac{3}{2}T_s$, e.g.; $|G_{2T_S}|^2$ will, as a function of τ, have values for τ near zero and τ near T_s—the latter being the range ambiguity, of course. The presence of the azimuth ambiguity is due to $|\tilde{H}(\frac{1}{2}n/\omega_s; \tau)^2|$ not being zero for $n = \pm 1$, of course. The term $n = 0$, with τ near zero, must produce the desired ambiguity function $|q_D|^2$. These are the three conditions that must be satisfied simultaneously if our problem has a solution.

In order to write out these conditions, we denote the signal modulation by

$$F(t) \equiv \sum_{-\infty}^{\infty} [F_1(t + \tfrac{1}{2}T_s - n2T_s) + F_2(t - \tfrac{1}{2}T_s - n2T_s)]. \tag{9.10}$$

Recall that we are assuming that the supports $(-\frac{1}{2}T_1, \frac{1}{2}T_1)$ and $(-\frac{1}{2}T_2, \frac{1}{2}T_2)$ of F_1 and F_2, respectively, are such that $2T_i \ll T_s$, for $i = 1, 2$.† Then, for τ

† The reason for this assumption is that, if it were not true, then the usual two-dimensional processing would be very greatly complicated, for otherwise the signal history of a given pattern could appear in part at the two delay extremes of the two-dimensional recorded signal—the processor would not be even quasi-invariant.

near zero,

$$\tilde{G}_{2T_s}(\tfrac{1}{2}n\omega_s\,;\tau) = \int e^{-\frac{1}{2}inw_st}[F_1(t+\tfrac{1}{2}T_s-\tau)\bar{F}_1(t+\tfrac{1}{2}T_s) + F_2(t-\tfrac{1}{2}T_s-\tau)\bar{F}_2(t-\tfrac{1}{2}T_s)]\,dt, \tag{9.11}$$

and, for τ near T_S,

$$\tilde{G}_{2T_s}(\tfrac{1}{2}n\omega_s\,;\tau) = \int e^{-\frac{1}{2}inw_s}[F_2(t+\tfrac{1}{2}T_s-\tau)\bar{F}_1(t+\tfrac{1}{2}T_s) + F_1(t-\tfrac{1}{2}T_s-\tau)\bar{F}_2(t-\tfrac{1}{2}T_s)]\,dt. \tag{9.12}$$

Condition for no range ambiguities. In order that there be no range ambiguities at $\tau = T_S$, we must have, by Eq. (9.9),

$$|\tilde{G}_{2T_s}(\tfrac{1}{2}n\omega_s\,;\tau)|^2 \equiv 0, \qquad \text{for} \quad n = 0, \pm 1.$$

We employ Eq. (9.12) and use Parseval's identity—much as we did in deriving Eq. (9.3a) from Eq. (9.3)—to write this condition as

$$\left|\frac{1}{2\pi}\int e^{-iu\tau}[\bar{\tilde{F}}_2(u-\tfrac{1}{2}n\omega_s)\tilde{F}_1(u) + (-1)^n\bar{\tilde{F}}_1(u-\tfrac{1}{2}n\omega_s)\tilde{F}_2(u)\,du]\right|^2 \equiv 0, \qquad n = 0, \pm 1, \quad \text{all } \tau.$$

But by the uniqueness of the Fourier transform, this implies that

$$\bar{\tilde{F}}_2(u-\tfrac{1}{2}n\omega_s)\tilde{F}_1(u) + (-1)^n\bar{\tilde{F}}_1(u-\tfrac{1}{2}n\omega_s)\tilde{F}_2(u) \equiv 0, \qquad \text{for } n = 0, \pm 1, \quad \text{all } u. \tag{9.13}$$

Condition for no azimuth ambiguities. In order that there be no azimuth ambiguities in the "τ near zero" strip, we must have, by Eq. (9.9)

$$|\tilde{G}_{2T_s}(\tfrac{1}{2}n\omega_s\,;\tau)|^2 \equiv 0, \qquad \text{for } n = \pm 1.$$

We employ Eq. (9.11), Parseval's identity, and the uniqueness of the Fourier transform to write this condition as

$$\bar{\tilde{F}}_1(u-\tfrac{1}{2}n\omega_s)\tilde{F}_1(u) - \bar{\tilde{F}}_2(u-\tfrac{1}{2}n\omega_s)\tilde{F}_2(u) \equiv 0, \qquad \text{for } n = \pm 1, \quad \text{all } u. \tag{9.14}$$

Condition for desired response. In order that the desired response be obtained in the "τ near zero" strip, we must have, by Eq. (9.9),

$$|\tilde{G}_{2T_s}(0;\tau)|^2\,|\tilde{H}(0;\mu)|^2 = |q_D(\tau,\mu)|^2.$$

That is,

$$\left|\frac{1}{2\pi}\int e^{-iu\tau}[|\tilde{F}_1(u)|^2 + |\tilde{F}_2(u)|^2]\,du\right|^2 \cdot \left|\int A(t-\mu)\bar{A}(t)\,dt\right|^2 = |q_D(\tau,\mu)|^2. \tag{9.15}$$

Impossibility of solution. We note that the required condition, Eq. (9.15), appears to contain a good deal of the difficulties of the general ambiguity problem mentioned at the beginning of this section. Fortunately, we shall not have to deal with it, except to note the perfectly obvious fact that $F_1 \equiv F_2 \equiv 0$ is not a permissible solution!

We shall now invoke the assumption that $T_i \ll T_S$, for $i = 1, 2$, to simplify the conditions, Eqs. (9.13) and (9.14). For, if the durations T_1 and T_2 of F_1 are very small relative to T_S, then $\tilde{F}_1$ and $\tilde{F}_2$ must be approximately constant over a distance $\frac{1}{2}\omega_S = \pi/T_S$.

With this result, we now obtain from Eq. (9.13) two equations: for $n = \pm 1$, we get

$$\bar{\tilde{F}}_2(u)\tilde{F}_1(u) - \bar{\tilde{F}}_1(u)\tilde{F}_2(u) = 0,$$

and for $n = 0$, we get

$$\bar{\tilde{F}}_2(u)\tilde{F}_1(u) + \bar{\tilde{F}}_1(u)\tilde{F}_2(u) = 0.$$

Summing these two equations we get the necessary condition

$$|\bar{\tilde{F}}_2(u)| \cdot |\bar{\tilde{F}}_1(u)| = 0. \tag{9.16}$$

From Eq. (9.14) we get the necessary condition

$$|\tilde{F}_1(u)|^2 = |\tilde{F}_2(u)|^2. \tag{9.17}$$

The only solution of Eqs. 9.16 *and* 9.17 *is* $|F_1| \equiv |F_2| \equiv 0$.

Thus we have shown that:

Property 9.6. "It is not possible to modify, in alternate periods, a periodic signal modulation to simultaneously: (1) remove the first range ambiguity, and (2) not introduce azimuth ambiguities, when the pulses are restricted to supports whose extent is very small relative to the period."

9.3.2. The Ambiguity Suppression Problem

Although not as precise and general as we would prefer, Prop. 9.6 does encourage a change in desideratum: rather than attempting to *avoid* ambiguities we should attempt to *suppress* the ambiguities. This is a very reasonable approach; it has been implicit in much of our discussion—especially

Chapter VIII—that suppression of spurious levels to a certain level should be sufficient, and this point of view is justified experimentally.

Thus, rather than attempting to require that $|G_{2T_s}(\frac{1}{2}n\,\omega_S;\,\tau)|^2$ be identically zero for certain n and τ-regions, we now attempt only to make it small. Retaining the practically important requirement that $T_1 T_2 \ll T_S$, we note that it is desirable to keep $|\tilde{F}_1|^2 \approx |\tilde{F}_2|^2$ in view of the condition, Eq. (9.14). If some spectral magnitude $|\tilde{F}|^2$ is known that produces the proper τ-behavior of $|q_D|^2$—as indeed it would be on the basis of the discussion of Chapter VIII, then it is also desirable to keep $|F_i|^2 \approx |\tilde{F}|^2$, for $i = 1, 2$, in view of the condition Eq. (9.15).

Finally, in view of Eq. (9.13), we would like F_1 and F_2 to have a low correlation. A pair of candidates easily comes to mind in the light of the discussion of Chapter II: F_1 and F_2 can be chosen to be linear FM signals with linear FM rates $+k$ and $-k$, respectively. Then (assume the large time-bandwidth-product case), their correlation has an approximately uniform magnitude, of relative level $1/\Delta T_1 \ll 1$ and of extent $2T_1(T_1 \equiv T_2)$. This is as good as one can hope.

It remains to establish the level of the azimuth ambiguities. It is easy to see that when

$$\bar{\tilde{F}}_1\left(u - \frac{\pi}{T_s}\right)\tilde{F}_1(u) \approx I_\Omega(u)\exp\left(\frac{i\pi u}{kT_s}\right),$$

$$\bar{\tilde{F}}_2\left(u - \frac{\pi}{T_s}\right)\tilde{F}_2(u) \approx I_\Omega(u)\exp\left(\frac{-i\pi u}{kT_s}\right),$$

their difference is approximately $I_\Omega(u)\cdot \pi u/kT_S$, and that the Fourier transform of their difference is bounded by $T_1\Omega/8T_S$, or relatively by

$$\frac{T_1}{8T_s} \ll 1.$$

Thus we have established that:

Property 9.7. "It is possible to modify alternate pulses of a periodic pulse train (of unmodified period T_S) in order to: (1) retain the desired ambiguity function near the origin; (2) keep the range ambiguities not greater than the reciprocal of the time-bandwidth-product of the pulse modulations; (3) keep the azimuth ambiguities not greater than $T_1/8T_s \ll 1$, where T_1 is the duration of (either) pulse modulation.

9.3.3. Further Remarks

The results of Sec. 9.3.2 can be generalized in various ways. For example, there are many ways to choose "quasi-orthogonal" modulations, in addition to the pair composed of opposite-slope linear FM. This is a well-known and much discussed communication theory problem.

We can simplify the appearance and sometimes the calculations of these forms and conditions, by recalling the "classical" range-Doppler radar ambiguity function

$$P_F(\tau, \mu) \equiv \int e^{-i\mu t} F(t-\tau)\bar{F}(t)\, dt,$$

associated with the modulation F and the "cross-ambiguity function"

$$P_{F_1F_2}(\tau, \mu) \equiv \int e^{-i\mu t} F_1(t-\tau)\bar{F}_2(t)\, dt.$$

These ambiguity functions, especially the first one, have been calculated for many modulations and appear scattered throughout the radar literature: one can therefore use such results to simplify the calculation of the SAR ambiguity function.

Using these ambiguity functions we can, for example, write the condition for no range ambiguities as

$$e^{\frac{1}{2}in\pi}[e^{-i\mu\tau}\bar{P}_{F_1F_2}(-\tau, -\mu) + (-1)^n P_{F_1F_2}(\tau, \mu)] = 0, \tag{9.18}$$

for $\mu = \frac{1}{2}n\omega_s/2$, $n = 0, \pm 1$, and the condition for no azimuth ambiguities as

$$e^{\frac{1}{2}in\pi}[P_{F_1}(\tau, \mu) - P_{F_2}(\tau, \mu)] = 0, \tag{9.19}$$

for $\mu = \frac{1}{2}n\omega_s, \quad n = \pm 1$.

Proposed "solutions" to the ambiguity problem are generally arrived at through intuitive or heuristic reasoning. Whatever the initial appeal such schemes may have, they are to be held in suspicion until their ambiguity function is calculated. This can be arduous, of course.

As an example, the idea of shifting the spectrum of F_2 so that it is disjoint with $\tilde{F}_1$ has the property of eliminating range ambiguities: F_1 and F_2 are obviously orthogonal, and $P_{F_1F_2} \equiv 0$—see Eq. (9.18). However, azimuth ambiguities result that are intolerable. A heuristic argument can be made that "explains" the presence of these ambiguities: see App. 9A. Here we calculate directly the left side of Eq. (9.19). Let $F_2 = F_1 e^{i\Omega t}$, where Ω is the bandwidth of F_1. Then,

$$P_{F_2}(\tau, \mu) = e^{-i\Omega\tau} P_{F_1}(\tau, \mu),$$

and the left side of Eq. (9.19) is

$$P_{F_1}(\tau, \mu)[1 - e^{i\Omega\tau}], \qquad \mu = \pm\tfrac{1}{2}\omega_s.$$

Now $P_{F_1}(\tau, \mu)$, as a function of τ, has appreciable value over the τ-interval $2\pi/\Omega^{-1}$, the range resolution interval. In this interval, $\exp i\Omega\tau$ assumes the value -1: *the azimuth ambiguity is of the same order as the central response!*

Rather complicated schemes have been proposed, which employ non-linear operations, usually by reasoning on the response of a single scatterer. Such schemes are likely to fail—or at any rate, should especially be checked—for multiple scatterers. Typically, false images are introduced by such schemes.

APPENDIX 9A: A DEFECTIVE AMBIGUITY "SOLUTION"

The idea of shifting the spectrum of the radiated wave-form from pulse to pulse is the "solution" that probably occurs first to most who consider the range ambiguity problem. We pointed out in Sec. 9.3.3 that shifting alternate pulses, in fact, can not avoid ambiguities. It is easy to give a quick argument showing the idea suspect.

Suppose that the radar set emits the pulse $F(t) \exp i\omega_1 t$, which is scattered by a point object at range delay τ and received as

$$F(t-\tau) \exp i[\omega_1(t-\tau) + \varphi(t; \lambda_1)],$$

where

$$\varphi(t; \lambda_1) \equiv (4\pi v^2/\lambda_1 R_m)(\tfrac{1}{2}t^2),$$

the usual SAR azimuth modulation. The next pulse emitted is

$$F(t) \exp i\omega_2 t,$$

and the received signal is

$$F(t-\tau) \exp i[\omega_2(t-\tau) + \varphi(t; \lambda_2)],$$

ignoring the delay T_s.

Now we assume that we want the system to operate just as if only one RF carrier were used, and that this means we must make the returned signals look the same. After heterodyning to a common carrier, the phase error remaining between the two returns is

$$\gamma = \tfrac{1}{2}(k_1 - k_2)t^2/2 - (\omega_1 - \omega_2)\tau.$$

The first term can be sufficiently well corrected—but not the second term, for we do not know τ. In fact, we know τ roughly to an accuracy of $1/\Omega_s$ sec, where Ω_s is the bandwidth (in Hertz) of F.† The phase error is therefore of the order of

$$2\pi \frac{(f_1 - f_2)}{\Omega_s} \quad \text{rad}.$$

Thus, if the RF carrier f_2 is shifted sufficiently that the spectra of the emitted waveforms do not overlap—$f_1 - f_2 \geq \Omega_s$—the resulting phase error is $\gamma \geq 2\pi$ rad from pulse to pulse. Referring to Chapter VI, such a phase error presumably cannot be tolerated.†

† These are the weakest parts of this heuristic argument. Attempting to improve it is not a fruitful direction to pursue.

Of course, to see what really happens when the RF frequency is shifted, one simply computes the ambiguity function $|q|^2$ for such a modulation. The effects are characterized, as they should be under the ML criterion for a point scatterer, by $|q|^2$—and one can ponder its form as one wishes.

COMMENTS AND REFERENCES

The first radar ambiguity problem, that for the "classical" range-Doppler radar that estimated the range and velocity of a point scatterer, was discussed in the well-known work of Woodward (1953). Since that time this particular ambiguity function has been studied at some length but still not completely satisfactorily; however, it is not directly relevant to SAR systems. See, however, the informative general remarks of Siebert (1956), and the summary in Cook and Bernfeld (1967).

A general setting of estimating parameters of radar targets from the maximum likelihood point of view was given by Kelly *et al.* (1960). The only published work on the ambiguity function of SAR systems is that of Harger (1965b), and Harger and Crimmins (1968).

X

System Design B: Phase Errors

10.1. INTRODUCTION

We arrive now, as a matter of course, at the final level of discussion of the SAR system design: attempting to take into account, a priori, the presence of phase errors which we have found (Chapter VI) to have such a deletrious effect upon the systems designed to cope with additive noise (Chapter IV).

Generally speaking, the criteria we used before carry over in a natural way, although the calculations are now more complex. The resolution, SNR_{max}, and least squares criteria are discussed first, assuming an 'LIV processor. We obtain only partial results with the first two criteria and show that we can gain little improvement under the third.

We next take up the ML criterion, which does not a priori assume a linear structure; and we find—with certain assumptions to gain the Gaussian distributions that ease the calculation—that, in fact, *nonlinear* structures are required. We devote special attention to the important problem of estimating the usually unknown center frequency of the azimuth modulation.

10.2. RESOLUTION AND SNR_{max} CRITERIA

We consider first the resolution criterion—energy in the desired resolution interval—that we have discussed at several points above. It turns out that this criterion and the SNR_{max} criterion—modified slightly here—lead to nearly the same optimum processor. These processors are each characterized by a Fredholm integral equation: we do not solve either but are able to extract some results of interest.

10.2.1. Resolution Criterion

We turn again to the discussion of resolution criteria. We have considered these criteria in the noise-free situation and with additive noise (Chapter IV) in order to synthesize the SAR system. The effects of phase errors, as measured by resolution criteria, have also been considered (Chapter VI). We now wish to acknowledge at the outset the presence of phase errors and to find the SAR system optimum under a resolution criterion. As before, we shall use relative energy in the desired resolution interval as the resolution measure, and the criterion is, of course, to maximize this relative energy. We shall restrict ourselves to linear, time-invariant systems for the usual reasons, keeping in mind that the SAR system processor will necessarily be quasi-invariant (Sec. 4.1), and we shall assume a phase-matched processor.

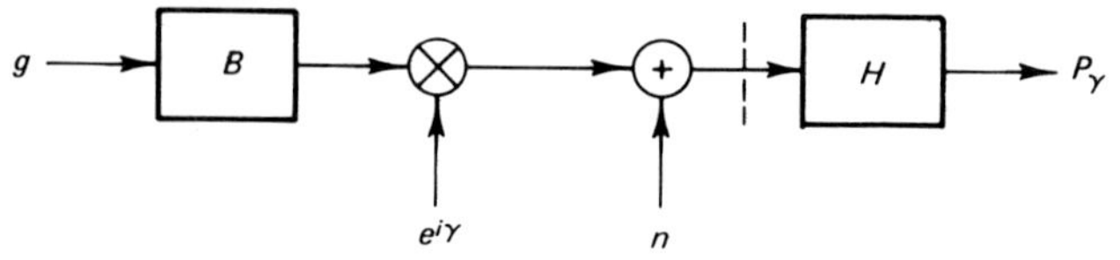

FIG. 10.1 Model for SAR system azimuth channel that includes phase errors.

Referring to Fig. 10.1 for the model of the SAR system azimuth channel, we have

$$p_\gamma = H * n + H * [e^{i\gamma}(B * g)]. \tag{10.1}$$

We ignore the additive noise explicitly in Eq. (10.1) while discussing resolution, but implicitly we keep in mind that, if the energy in the desired resolution interval becomes small, then the SNR will be small also. Of course we set $g = \delta$, a "point" scatterer, in Eq. (10.1), to get the system impulse response. As in Chapter VI, we note that B_m, for the usual SAR system design, is such that $B_m(y - x) \approx B_m(y)$ for $x \in I$, the desired resolution interval: this affords the considerable simplification that $p_\gamma(x)$ can be represented as a Fourier transform: if $H_m \equiv |H|$, and $\arg H \equiv -\arg B$,

$$p_\gamma(x) = \int dy\, B_m(y) H_m(y) \exp(i\gamma(y - x) - ikxy). \tag{10.1a}$$

(We check that if $H \equiv \bar{B}$, then $|p_\gamma|^2 = |q_\gamma|^2$ from Eq. (6.2).

The average energy in the azimuth resolution interval I is directly calculated to be

$$\begin{aligned} E_{I\gamma} &\equiv E\left\{\int_I |p_\gamma|^2\, dx\right\} \\ &= \int dy\, B_{my} H_{my} \int d\zeta\, B_{m\zeta} H_{m\zeta} \frac{\sin[\frac{1}{2}kI(y - \zeta)]}{\frac{1}{2}k(y - \zeta)}\, \phi_{\Delta\gamma}(1; y - \zeta). \end{aligned} \tag{10.2}$$

Compare the analogous equation in Sec. 6.2.2.

The energy $E_{I\gamma}$ is to be maximized. To have the problem well set, we need to maximize it relatively. A good choice to measure $E_{I\gamma}$ against is the energy E_{I0} in the resolution interval I, in the absence of phase errors: but in a well-designed system, this very closely approaches

$$E_T \equiv \frac{2\pi}{k} \int B_m{}^2(y) H_m{}^2(y)\, dy, \qquad \text{(refer to Chapter VIII).} \qquad (10.3)$$

Therefore, we wish to maximize the ratio $E_{I\gamma}/E_T$ as defined by Eqs. (10.2) and (10.3).

We note that B_m and H_m enter only as the product $B_m H_m$; we therefore could consider the problem of maximizing $E_{I\gamma}/E_T$ with respect to $C_m \equiv B_m H_m$. We do not do this for the following reasons:

1. The optimum C_m of course, depends on the nature of the phase error process $\{\gamma\}$ via the characteristic function of its increments $\Delta\gamma$. This phase error process is usually different for different mapping runs: the portion of the atmosphere of interest may be clear and calm, it may be a turbulent cold front, etc. If B_m is varied as C_m varies, changes would be frequent.
2. Changing B_m requires changing the far-field antenna pattern: this is not easy in practice.
3. Changes in C_m can be accomplished by the practically much easier variation of the processor H, for the usual B_m.

We, therefore, will regard H_m as the variable for the optimization. We shall choose B_m as we did for the phase error free situation: if B_m is to be fixed, this is the only reasonable choice generally, for, not infrequently, conditions are such that phase errors are very small in well-designed systems—e.g., on "calm" days.

With this choice on B_m we can make some simplifications that enable us to see in part the nature of the solution. For as we noted in Sec. 6.2.2, $B_m(x)$ may then be approximated by a constant for $x \in (-\frac{1}{2}X, +\frac{1}{2}X)$, and zero elsewhere. The variational problem is

$$\max_{H_m \in \mathscr{L}^2} \frac{(H_m, KH_m)}{\|H_m\|^2},$$

where $(\cdot, \cdot)$ and $\|\cdot\|$ are the inner product and norm on $\mathscr{L}_2(-\frac{1}{2}X, \frac{1}{2}X)$ and the operator

$$Kf(y) \equiv \int_{-\frac{1}{2}X}^{\frac{1}{2}X} \left\{\frac{\sin[\frac{1}{2}kI(y-\zeta)]}{(y-\zeta)} \phi_{\Delta\gamma}(1, y-\zeta)\right\} f(\zeta)\, d\zeta. \qquad (10.4)$$

Since K is linear, completely continuous, and symmetric, we have the same type of variational problem encountered at several other points of our discussion (Chapters IV and VIII). We know a solution exists: the maximum is the largest positive characteristic λ_o^+ value of the integral equation

$$\lambda f = Kf, \tag{10.5}$$

and H_m is to be taken as the associated characteristic function f_o. (If λ_o^+ has nonunit multiplicity, then H_m can be taken as any linear combination of the associated characteristic functions—that is, H_m may be taken as any function in the subspace of λ_o^+.) Of course, Eq. (10.5) can generally be solved only approximately.

We observe first that, if $\phi_{\Delta\gamma}(1, y - \zeta) \approx 1$ for n, $\zeta \in X$, then once again H_m will be a member of the prolate spheroidal wave function family. That is, as we have repeatedly observed, if the phase error process $\{\gamma_y\}$ does not fluctuate over the aperture X, it is of no consequence here.

A bound on λ. If we write our maximization problem in the form

$$\max_{H_m} \lambda, \qquad \lambda \equiv \frac{(H_m, KH_m)}{\|H_m\|^2},$$

then as

$$(H_m, KH_m) \le \|H_m\| \cdot \|KH_m\| \le \|H_m\| \cdot \|K\| \cdot \|H_m\|,$$

we have $\lambda \le \|K\|$, the norm of the operator K. But $\|K\| \le \|k\|$, the L_2 norm of the kernel k of K over the square $X \times X$; since k is a function of differences —see Eq. (10.4)—we have

$$\|k\|^2 = \left(\frac{kT}{2\pi}\right)^2 \|\phi_{\Delta\gamma}\|^2.$$

Finally, we have

$$\lambda \le \sqrt{2}\,\frac{kIX}{2\pi}\left(\frac{\int_X |\phi_{\Delta\gamma}|^2}{X}\right)^{1/2}. \tag{10.6}$$

Now $kIX/2\pi$ is of the order of unity: therefore, if, e.g., the support of $\phi_{\Delta\gamma}$ becomes much less than X—recall $|\phi_{\Delta\gamma}| \le 1$—$\lambda$ is necessarily less than 1. In other words, when the essential support of $|\phi_{\Delta\gamma}|$ becomes small with respect to X, the percent energy in the resolution interval I falls off, no matter how we choose H_m.

The nature of $\phi_{\Delta\gamma}$. When the phase errors are sufficiently small that $\exp i\gamma \approx 1 + i\gamma - \frac{1}{2}\gamma^2$ with large probability then, as we have noted already,

$$\phi_{\Delta\gamma}(1; y) \approx 1 - \sigma^2[1 - \rho(y)].$$

Note that no specific distributions need be assumed.

If the phase errors are not small, we generally must assume some type of distributions in order to say more. Suppose that $\{\gamma_y\}$ is a Gaussian random process; then

$$\phi_{\Delta\gamma}(1; y) = \exp -\sigma^2[1 - \rho(y)].$$

If the phase errors are large, $\sigma^2 \gg 1$, then it is possible to argue that

$$\phi_{\Delta\gamma}(1; y) \approx \exp -\tfrac{1}{2}\sigma^2\Omega^2 y^2,$$

where Ω is the bandwidth (measured by radius of gyration, say) of the spectral density of $\{\gamma\}$.

These forms can be used in Eq. (10.6) to bound the ratio $\lambda \equiv E_I/E_T$.

Further comment. The problem we have formulated, based on the ratio E_I/E_T, involves a severe demand when phase errors are large. In the latter situation, it is still possible that the system would be useful when operated with a reduced resolution requirement. That is, when the phase errors are large, it can be an attractive procedure to give up trying for the resolution I, achievable in the absence of phase errors. For example, intuitively, if we pick H_m with an extent over which exp $i\gamma$ changes little (with large probability), then we would have a good system—with resolution about the reciprocal of the extent of H_m, and SNR reduced by the ratio of the extent of H_m divided by the aperture X.

As we said, there appear to be no formal results in this direction, so we shall drop the discussion of resolution at this point. However, when we take up the maximum likelihood procedure (Sec. 10.4.1) we shall find that the processor structure will consist of, in part, a processing over a reduced aperture that implicitly reduces the resolution.

10.2.2 SNR_{max} Criterion

We consider now the maximum signal-to-noise ratio criterion, interpreted for the situation when phase errors are present. We use the model of Fig. 10.1, whose output is given by Eq. (10.1) when the input is a point-scattering object. We have pointed out (Sec. 4.3.1) that this criterion is suitable when the presence or absence of an object is to be determined, and the details of the image—presumed small relative to system resolution—are not of interest, except perhaps for its geometrical location. Then it is reasonable to call the first term of Eq. (10.1) "noise" and the second term "signal." For a point-scattering object at relative coordinates (0, 0), we then have, reasonably,

$$SNR_\gamma \equiv \frac{E\{|H * (e^{i\gamma}B)|^2\}|_{(o,o)}}{E\{|H * n|^2\}}. \qquad (10.7)$$

Of course, we easily calculate—as we have several times already—

$$E\{|H * n|^2\} = \frac{N_o}{2\pi} \int |H(\omega)|^2 \, d\omega, \tag{10.8}$$

for white (spectral density amplitude N_o) additive noise, and

$$E\{|H * (e^{i\gamma}B)|^2\}|_{(o,o)} = \int dy \, B_m(y) H_m(y) \cdot \int d\zeta \, B_m(\zeta) H_m(\zeta) \phi_{\Delta\gamma}(1; y - \zeta). \tag{10.9}$$

Compare Eqs. (10.9) and (10.2).

Maximizing SNR_γ with respect to H_m is therefore the same type of problem we encountered with the energy resolution criterion in Sec. 10.2.1: the only difference is that the kernel of the relevant operator—compare Eq. (10.4)—is now just $\phi_{\Delta\gamma}(1; y - \zeta)$. Remarks of the same general nature therefore carry over from that problem to this one.

10.3. LMVE WITH PHASE ERRORS

We now take up the least squares criterion and find the best linear filter. We will assume, for reasons discussed earlier (Sec. 6.3), that the phase errors are small and of relatively broad bandwidth. More general remarks made in Secs. 4.2.2 and 4.3.2 remain pertinent.

We take up where we left off in Sec. 6.3; reference to Fig. 6.6 will recall the details of the problem and we examine Eq. (6.14) now in order to find the optimum filter h_{opt}:

$$\min_a \mathrm{MSE}(ah) = \sigma_g{}^2 - \frac{(1 - \sigma^2)}{2\pi} \frac{|\int \bar{\tilde{h}} \psi_1 \, d\omega|^2}{\int |\tilde{h}|^2 \psi_2 \, d\omega}, \tag{10.10}$$

where

$$\psi_1 \equiv \bar{\tilde{B}} P_g$$

and

$$\psi_2 \equiv P_n + |\tilde{B}|^2 P_g + \frac{1}{2\pi} P_\gamma * (|\tilde{B}|^2 P_g).$$

We proceed in the usual way to minimize this MSE by maximizing the second term on the right side of Eq. (10.10): by the Schwarz inequality,

$$\left| \int \bar{\tilde{h}} \psi_1 \right|^2 \le \int |\tilde{h}|^2 \psi_2 \cdot \int \frac{|\psi_1|^2}{\psi_2},$$

with equality, if and only if, $h = \psi_1/\psi_2$ within a complex constant scale. Thus, within such a scale,

$$\tilde{h}_{\text{opt}} = \frac{\bar{\tilde{B}} P_g}{P_n + |\tilde{B}|^2 P_g + \dfrac{1}{2\pi} P_\gamma * (|\tilde{B}|^2 P_g)} \tag{10.11}$$

and

$$\min_{a,h} \text{MSE}(ah) = \sigma_g^{\,2} - \frac{(1-\sigma^2)}{2\pi} \int \frac{|\tilde{B}|^2 P_g^{\,2}}{P_n + |\tilde{B}|^2 P_g + \dfrac{1}{2\pi} P_\gamma * (|\tilde{B}|^2 P_g)}. \tag{10.12}$$

We may make remarks about Eqs. (10.11) and (10.12) very similar to those made about Eq. (6.14) because $P_\gamma * (|\tilde{B}|^2 P_g)$ can usually be replaced to a good approximation by its value at $\omega = 0$. Thus, *so far as linear minimum variance estimation is concerned, the (multiplicative) phase error can be replaced by an equivalent additive, independent (of g and n) noise process,* whose spectral density, $(2\pi)^{-1} P_\gamma * (|\tilde{B}|^2 P_g)$, can usually be taken as white over the spectral band of interest—namely, the spectral band where the major support of $|\tilde{B}|^2 P_g$ appears.

Of course, more a priori information—namely P_g—is required—e.g., must be measured—in order to set up the filter h_{opt} of Eq. (10.11). We essentially need to know the variance and the bandwidth of the phase error process. Usually, of course, P_n is white over the spectral band of interest, and so, if the phase errors are small, the optimum filter with phase errors, Eq. (10.11), does not differ greatly from the phase-error-free filter of, e.g., Eq. (6.15).

Since this added information comes at some cost, it is informative to know what improvement is actually obtained. Using the same definitions of I_j, for $j = 1, 2, 3$, introduced in Sec. 6.3, we can write Eq. (10.12) in the form

$$\min_{a,h} \text{MSE}(ah) = \text{MSE}\,|_{\gamma=0} + \sigma^2 I_1 + (1-\sigma^2) I_3.$$

We compare this now with

$$\min_{a} \text{MSE}(ah)\,|_{h=h_*} = \text{MSE}\,|_{\gamma=0} + \sigma^2 I_1 + I_3,$$

where h_* is given by Eq. (6.15)—that is, it is the LMVE filter optimum when $\gamma \equiv 0$—and we have adapted Eq. (6.16). Thus

$$\min_{a} \text{MSE}(ah)\,|_{h=h_*} - \min_{a,h} \text{MSE}(ah) = \sigma^2 I_3, \tag{10.13}$$

where we recall

$$I_3 \equiv \frac{1}{2\pi} \int \frac{|\tilde{B}|^2 P_g [P_\gamma * (|\tilde{B}|^2 P_g)]}{(P_n + |\tilde{B}|^2 P_g)^2}.$$

Thus, the improvement of the optimum filter, relative to the filter optimum for $\gamma \equiv 0$, *is of the order of* σ^4.

We conclude, therefore, that for broadband, small phase errors, the extra complication of the optimum LMVE filter, Eq. (10.11), relative to the filter, Eq. (6.15), is probably not worth the benefit derived.

10.3.1. Design of $|\tilde{B}|$

It is clear that, at least in principle, we can consider the further problem of the form Eq. (10.12), for the mean-square error, by variation over a set $\{|\tilde{B}|\}$ of possible azimuth modulation spectral magnitudes. (We note that min MSE(*ah*) does not depend on the phase of the azimuth modulation "pre-filter.") This is not an easy problem, as $|B|$ enters in part via the convolution $P_\gamma * (|\tilde{B}|^2 P_g)$. Also, if we wished to be precise, we would have to note that $|\tilde{B}| \approx |E_{oh}|^2$, the two-way voltage far-field azimuth pattern, is actually the square of the Fourier transform of a function of compact support: we shall ignore this, adopting the position that any solution we find here would then have to be approximated by such a function. We shall impose an energy restraint, $\int |\tilde{B}|^2 =$ constant.

In order to solve this problem at all, we assume that, though the phase error process $\{\gamma\}$ is of broad bandwidth relative to $2\pi/X$, where X is the synthetic aperture length, it still fluctuates slowly over an azimuth resolution distance. This is a reasonable practical assumption, e.g., for aircraft vehicle motion. Then it appears reasonable to assume that the optimum $|\tilde{B}|$ will be broad with respect to P_γ and hence that

$$P_\gamma * (|\tilde{B}|^2 P_g) \approx \sigma^2 |\tilde{B}|^2 P_g, \tag{10.14}$$

unless restraints on $|\tilde{B}|$ force it to be otherwise. We proceed with this assumption that allows us to eliminate the convolution appearing in Eq. (10.12): whether or not any particular solution satisfies the assumption can always be checked.

Thus, we consider the variational problem

$$\min_{|\tilde{B}|} \min_{a,h} \mathrm{MSE}(ah), \qquad \text{subject to} \quad \int |\tilde{B}|^2 = \text{constant},$$

where

$$\min_{a,h} \mathrm{MSE}(ah) \equiv \frac{1}{2\pi} \int \frac{P_g(P_n + \sigma^2 |\tilde{B}|^2 P_g)}{P_n + (1 + \sigma^2) |\tilde{B}|^2 P_g} \, d\omega.$$

We have simplified the last formula, employing Eq. (10.14) in Eq. (10.12). This problem is fairly easily solved by the method of Lagrange multipliers: we find

$$|\tilde{B}|^2 = \begin{cases} \max\left[\dfrac{1}{1+\sigma^2}\left(\dfrac{(P_n)^{1/2}}{\lambda} - \dfrac{P_n}{P_g}\right), 0\right], & \omega \ni P_g(\omega) \neq 0, \\ 0, \quad \omega \ni P_g(\omega) = 0, \quad \text{and} \quad \lambda \neq 0. & \end{cases} \tag{10.15}$$

Thus in this case, $|\tilde{B}|^2$ *is unchanged from the case* $\gamma \equiv 0$ [refer to Eq. (6.7)] except for an uninteresting scale.

The general conclusion that we may hazard, based on the results of this section, is that, so far as linear minimum variance estimation is concerned, small phase errors that fluctuate over the synthetic aperture distance do not merit changing the SAR system design optimized for additive noise alone. This is an important conclusion, since the system complexity would be considerably increased in order to include the capability of measuring the relevant parameters of the phase error process.

10.4. MAXIMUM LIKELIHOOD CRITERIA

In the quest for processing structures that can partially offset the degradations of phase errors, we turn again to the maximum likelihood criterion. Its advantages and disadvantages have already been discussed.

The first problem posed is much like before: we wish to detect and estimate the amplitude and geometrical position of a point-scattering object. When we posed this problem with an additive, Gaussian noise model we found an already familiar *linear* structure which, therefore, minimized importance of the point-scattering assumption in our application here. Now, however, when we include phase errors along with the additive Gaussian and make a certain other appeal to a Gaussian assumption (Sec. 10.4.1), we get a *nonlinear* processing structure. Further, the physical realization of this processor by an optical system is quite complicated, although it can simplify somewhat in certain cases.

10.4.1. Point Scatterer Estimation

Referring again to the azimuth signal channel model of Fig. 10.1, the signal/data available for processing is

$$Z(x) \equiv e^{i\gamma(x)}B(x) + N(x), \qquad x \in X,$$

assuming a point scattering object. We assume the additive noise $\{N_x\}$ is a white, Gaussian process (of mean zero and constant spectral density N_o), but even if $\{\gamma_x\}$ is a Gaussian process, $\{Z_x\}$ will not be a Gaussian process: this makes finding the likelihood functional more difficult.

If, however, we fix the phase error γ_x, the conditional likelihood functional is well known: for then $\{Z\}$ is a Gaussian process, and the likelihood functional we have already stated in Sec. 4.3.3 is:

$$L(Z \mid \gamma) = \exp\left\{\frac{2}{N_o} \operatorname{Re} \int_x \bar{Z} e^{i\gamma} B - \frac{1}{N_o} \int_x |Be^{i\gamma}|^2\right\}.$$

If we now average $L(Z \mid \gamma)$ (over the probability space of γ) we obtain the desired likelihood functional. In order to carry out this average, it is convenient to make one of the two following assumptions:

1. The phase errors are sufficiently small that $\exp i\gamma \approx 1 + i\gamma$, and $\{\gamma_x\}$ is a Gaussian process.
2. The bandwidth of $\{\gamma_x\}$ is large, relative to $1/X$, and by a (hopeful) appeal to the Central Limit Theorem, the exponent of $L(Z \mid \gamma)$ is a Gaussian random variable.

For then we have only the well-known calculation $E\{\exp \alpha\}$, where α is a Gaussian random variable.

When the phase errors are small, we find

$$L(Z) \approx \frac{2}{N_o} \int_x \bar{Z}B - \frac{1}{N_o} \int_x |B|^2 - \frac{\sigma^2}{N_o^{\,2}} \operatorname{Re}\left\{\int_x \bar{Z}_X B_X \int_X \bar{Z}_y B_y \rho(x-y)\right\}$$

$$+ \frac{\sigma^2}{N_o^{\,2}} \int \bar{Z}_x B_x \int Z_y \bar{B}_y \rho(x-y). \tag{10.16}$$

When the phase errors are of broadbandwidth, we find

$$L(Z) \approx \frac{2\exp(-\frac{1}{2}\sigma^2)}{N_o} \operatorname{Re} \int_x \bar{Z}B - \frac{1}{N_o} \int_x |B|^2$$

$$+ \frac{\exp(-\sigma^2)}{N_o^{\,2}} \operatorname{Re}\left(\int_x \bar{Z}_x B_x \int_x \bar{Z}_y B_y \{\exp[-\sigma^2 \rho(x-y)] - 1\}\right)$$

$$+ \frac{\exp(-\sigma^2)}{N_o^{\,2}} \int_x \bar{Z}_x B_x \int_x Z_y \bar{B}_y \{\exp[\sigma^2 \rho(x-y)] - 1\}. \tag{10.17}$$

Equations (10.16) and (10.17) give, in two special cases, the optimum processing: clearly the structure is nonlinear, involving quadratic functionals of the data. We verify that as $\sigma^2 \to 0$, both forms revert to the likelihood functional for additive, Gaussian noise only—the first two terms of Eq. (10.16)—which we have called the "matched filter" processor.

The nature of the quadratic functionals can be shown to have considerable appeal. For example, consider the last term of both Eqs. (10.16) and (10.17). The kernel—

$$\psi(x) \equiv \exp(-\sigma^2)\{\exp[-\sigma^2 \rho(x)] - 1\} \tag{10.18}$$

in Eq. (10.16b), e.g.—tends to have value unity when $\rho(x) \approx 1$, and small value elsewhere, very small as σ^2 increases. Thus the *interior* integral will sum $\bar{Z}B = (\bar{B} \exp i\gamma + n)B$ over regions where, with good probability, (1) the phase γ is constant, and (2) $\gamma(x) = \gamma(y)$. Thus $\bar{Z}_x$ times the inner integral tends to be independent of the phase error, ignoring cross-products with the noise n, over a region. Thus we may say that this quadratic functional *coherently* integrates/sums over regions that, on the average, have equal phase errors, and *incoherently* integrates/sums the contribution for all such regions. It has been called a *mixed integrator* processor. The other quadratic functional can be similarly interpreted.

The size of the region over which the coherent integration is carried out is determined by the kernel of the quadratic functional: it depends on both the variance and the bandwidth (or correlation distance) of the phase error process —as we are accustomed to by now. This provides insight into the unsolved problem of Sec. 10.2: we could choose H_m now to be ψ. Of course the processor we have found here is considerably more complicated than a simple linear filter.

We will not pursue the discussion of this processor further here. The present discussion has provided a structure that is very interesting but has not been applied to SAR systems as yet. However, the idea of mixed integration has been applied, and these results give a rational way of picking H_m (see Sec. 10.2) when only a linear processor is to be used. The effects of the nonlinear nature of the processor in a dense and distributed scattering situation have not been investigated. We note finally that the mixed integrator has a very simple realization by a coherent optical system.

10.4.2. Center Frequency Estimation

It is difficult in practice to position, and keep positioned, the physical antenna beam, especially because of spurious vehicle motion. This fact has its most important consequence, as usual, in the azimuth channel of the SAR system. In practice, the azimuth boresight angle will be slowly varying, introducing in part a linear phase error as we will see immediately below. Furthermore, this angle is often unknown to the required precision. For example, for an orbital vehicle in an inclined and eccentric orbit, perturbed by various torques, the azimuth angle of the antenna boresight will not be known; further because of the changing nature of the spurious torques and, in fact, the changing velocity vector of the earth's illuminated area with vehicle latitude change, this unknown angle will be a function of time.

As we remarked in Chapter II, one usually wants the azimuth boresight to lie in the zero-Doppler direction. Although even for a satellite vehicle this direction can be computed and supplied to the vehicle attitude control system, it may be impractical or impossible to require the control system to perform this operation. This, in fact, is a complicated system question that interacts at many levels with the overall vehicle system design.

We remind ourselves that the importance of knowing the azimuth boresight angle relative to the zero-Doppler direction is so that we can keep the spectra of the ensemble of signals in the passband of the SAR system processor. As we have remarked at various points, this bandwidth is expensive by any of several measures, and excess to allow for varying offset frequencies of the data is usually not possible.

It is, of course, obvious that the data ensemble *itself* contains implicitly the information we need to estimate the unknown center frequency caused by the unknown azimuth boresight angle. Having estimated this center frequency, we could now do one of several things: (1) control the frequency of a local oscillator to center the spectrum of the data ensemble in the system passband, (2) electronically steer the antenna in azimuth to the desired azimuth angle, or (3) physically move the antenna by, e.g., moving the vehicle itself. Each of these methods has advantages and disadvantages of a fairly obvious nature.

The first method is relatively simple to implement, but some discretion is required in its use: it must not be allowed to introduce phase errors into the azimuth histories. In fact, it is easy to show that if such a scheme is used to "track" the change in center frequency change due to latitude change of a typical orbital vehicle, then a small but not negligible quadratic phase error will be introduced into the azimuth histories.

We shall content ourselves here with deriving the effect of an offset point angle—which contains a subtlety—with showing how the unknown center frequency is estimated, and by pointing out a feedback control system that accomplishes the correction.

The effect of an offset angle. As we discussed in Sec. 2.4, the principal effect of having the antenna beam pointing at an azimuth angle α away from the zero-Doppler angle is to produce an ensemble of signals on a carrier, or offset, frequency associated with the angle. There is a more subtle effect that is of sufficient interest to point out: we therefore quickly rederive the form of the scattered signal with now an offset angle in azimuth.

Referring to Fig. 10.2, we calculate the phase approximation much as we did originally (Chapter II), except that we factor $R*^2 \equiv (R_o^2 + x_o^2)$ out of the expression

$$R = [R_o^2 + (x_o + \xi)^2]^{1/2}, \qquad x \equiv x_o + \xi, \tag{10.19}$$

before using the binomial expansion. The resulting approximation to the returned signal is

$$S(t) = \int B_h(\xi)\{\phi(x_o + \xi - vt)\exp[-i(2k\xi \sin \alpha)]\}\, d\xi \cdot \exp[i\omega_o(t - t_*)], \tag{10.20}$$

for the strip of reflectivity at delay $\tau_* \equiv 2R_*/v$; in this equation $k = 2\pi/\lambda$.

We note that the returned signal S is on a mean frequency $\omega_o + 2kv \sin\alpha$ instead of ω_o. (To see this quickly, set $\phi(x) \equiv \delta(x)$, and recall that B_h is generally symmetric about zero.) We note second that a *different* spatial frequency spectrum of ϕ is observed ($\alpha \neq 0$): e.g., if $\lambda = 0.1$ ft, the observed spectrum is centered on about 100 sin α cycles/ft.

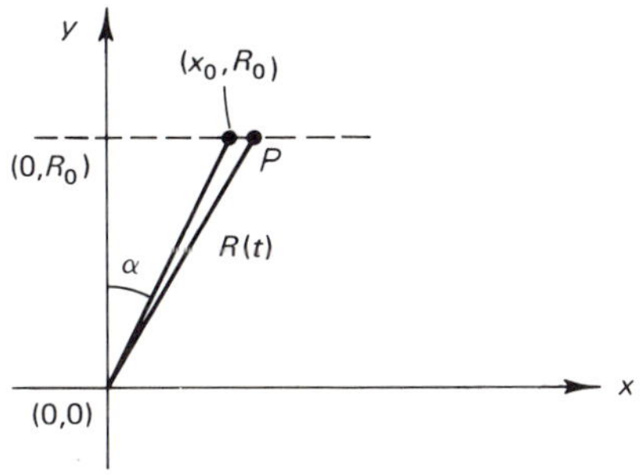

FIG. 10.2 SAR system geometry with antenna boresight offset azimuth angle α from zero-Doppler direction.

The reason why the second observation is of less practical importance is seen when we think again of the basic geometry of the SAR system (Fig. 1.1). Because of the generally inclined angle of the illumination, the SAR system receives energy from that portion of the terrain that has high spatial frequency, in order to backscatter energy to the radar set. At these spatial frequencies, one can expect that a small shift in observed spectrum would not change the nature of the image.

Therefore, ignoring the second effect, we have that an offset angle α causes a linear phase error

$$\gamma(t) = \frac{4\pi v}{\lambda}\alpha t, \qquad \alpha \ll 1,$$

$$\equiv \alpha_1 t, \tag{10.21}$$

in the time domain. We must, of course, operate on the temporal, "one-dimensional" signal and correct it before we record it as a spatial, two-dimensional signal.

Estimating the center frequency. The data received is a random process whose spectral density is known to us quite well, being determined by the two-way antenna pattern; if we assume that the terrain process has a "white" spectrum over the band of interest, then we know the spectrum to within its amplitude:

$$P_z(\omega) = P_\varphi(0) \cdot |\tilde{B}(\omega - \alpha_1)|^2 + P_n \equiv P(\omega - \alpha_1) + P_n. \tag{10.22}$$

Usually, P_n is white over the band of interest also.

In writing Eq. (10.22), we have ignored the sampling by effectively considering just one alias. In practice, this data could be made available in an auxiliary channel that first heterodynes the data down to an offset frequency and then filters the data to preserve one (usually displaced) alias.

There are several ways to argue to a particularly reasonable procedure to estimate α_1. We choose the method of maximum likelihood once again, for the same reasons as before. Its particular drawback is the necessity of assuming Gaussian statistics in order to proceed. We also assume, very reasonably, that the process correlation distance is much shorter than the time extent over which we collect data to estimate α_1.

The logarithm of the likelihood functional, i.e., optimum processor, can be shown to be

$$\mathrm{L}\{S\} = -\frac{T_c}{4\pi}\int \log\left[1 + \frac{P(\omega - \alpha_1)}{N_o}\right] + \frac{1}{4\pi}\int \frac{|\tilde{S}_{T_c}|^2}{N_o} \frac{P(\omega - \alpha_1)\, d\omega}{N_o + P(\omega - \alpha_1)}, \tag{10.23}$$

where $\tilde{S}_{Tc}$ is the spectrum of the T_c-length piece of data S_t. Since the first term of $\mathrm{L}\{S\}$ is independent of α_1 we ignore it. Thus, the maximum likelihood estimate $\hat{\alpha}_1$ of α_1 is that value(s) of α_1 that maximizes the form

$$I(\alpha_1) \equiv \frac{1}{4\pi}\int \frac{|\tilde{S}_{T_c}|^2}{N_o} \frac{P(\omega - \alpha_1)}{N_o + P(\omega - \alpha_1)}\, d\omega. \tag{10.24}$$

Equation (10.24) describes an intuitively reasonable procedure: (1) we calculate essentially the "periodogram" $|\tilde{S}_{Tc}|^2$ of the data as our best available guess of the spectral density of the data; (2) we form the inner product of the periodogram and a filter transfer function whose center frequency is variable; (3) the inner product is maximized by choice of α_1.

We note that when $P(\omega - \alpha_1) \gg N_o$, the filter transfer function becomes essentially unity over the essential support of $P(\omega - \alpha_1)$ and *we do not need to know the amplitude of* $P(\omega)$ *in order to realize the optimum procedure.* This condition usually is true in practice and affords an important practical simplification. One can show that, in any case, the MLE's of $\hat{\alpha}_1$ and spectral amplitude are uncorrelated.

One can also show that, as T_c becomes very large relative to the azimuth resolution of the system, the variance of the MLE α_1 approaches the least possible:

$$\mathrm{Var}\{\hat{\alpha}_1\} \to \left\{\frac{T_c}{4\pi}\int \left[\frac{P'(\omega)}{N_o + P(\omega)}\right]^2 d\omega\right\}^{-1}. \tag{10.25}$$

Thus, e.g., the larger T_c the better: in the SAR system this is limited by the slow variation of α_1 itself.

We can use Eq. (10.25) to establish how much data should be processed in order to get a satisfactory estimate of center frequency. To get the essential idea, suppose that

$$P(\omega) \approx \begin{cases} P_o \dfrac{\exp[-\frac{1}{2}(\omega^2/\Omega^2)}{(2\pi)^{1/2}\,\Omega}, & |\omega| < \Omega, \\ 0, & |\omega| > \Omega, \end{cases}$$

and that

$$P(\omega) \gg N_o, \qquad |\omega| < \Omega.$$

Then

$$\frac{P'(\omega)}{P(\omega)} \approx \frac{-\omega}{\Omega^2}, \qquad |\omega| < \Omega,$$

and, integrating in Eq. (10.25),

$$\mathrm{Var}\{\hat{\alpha}_1\} \approx \frac{\Delta}{T_c}. \tag{10.26}$$

That is, the variance of the estimate of center frequency improves with increasing data collection time T_c—reasonably—and decreases with increased bandwidth.

As a rough rule of thumb, we would usually like the center frequency to be in error no more than a few percent of the signal bandwidth, in order to avoid the spurious effects mentioned in Chapter II and elsewhere. Thus, say we would like the standard deviation

$$\left(\frac{\Delta}{T_c}\right)^{1/2} < \frac{\Delta}{\theta},$$

where, typically, $10 \leq \theta \leq 10^2$. Thus, we must have

$$\Delta T_c > \theta^2, \qquad 10 \leq \theta \leq 10^2. \tag{10.27}$$

As an example, in a typical orbital SAR system one might have $\Delta T \approx 10^3$: then, stating the requirement, Eq. (10.27), in other words, if we choose $\theta = 10$, one should collect data over one-tenth the synthetic aperture history length to get a satisfactory estimate of center frequency.

We can also conclude from Eq. (10.27) that the center frequency must not change appreciably in a time less than θ^2/Δ sec, else it in fact cannot be estimated—by this system, at any rate.

Correcting the center frequency. The necessary condition for a maximum of Eq. (10.24) is found, of course, by differentiating with respect to α_1 and equating with zero: this "likelihood equation" can be written

$$I_{\alpha_1}(\alpha_1) \equiv \int \frac{|\tilde{S}_{T_c}(\omega + \alpha_1)|^2}{T_c} \cdot \frac{P'(\omega)}{[N_o + P(\omega)]^2}\, d\omega = 0. \tag{10.28}$$

This equation suggests, with more precision, a commonly used technique to estimate and correct the center frequency.† The incoming data is placed on a controllable offset frequency $\omega_1 + \alpha_1$ by a local oscillator: α_1 will be varied, and ω_1 is sufficiently large to ensure a bandpass situation. Then $\tilde{S}(\omega + \alpha_1)$ is the spectrum of the complex signal.

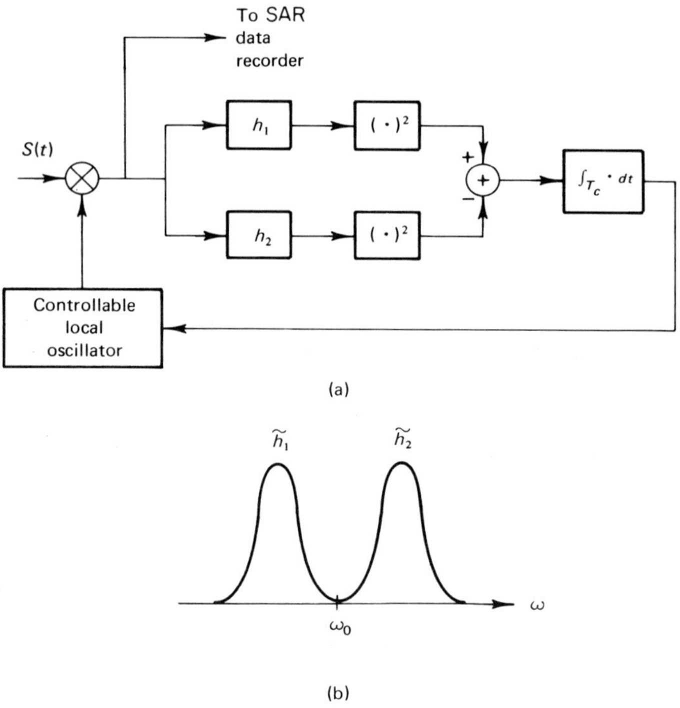

FIG. 10.3 (a) The feedback control circuit suggested by the likelihood equation for estimating and correcting the unknown center frequency of the data. (b) The transfer functions $\tilde{h}_1$ and $\tilde{h}_2$ of the prefilters h_1 and h_2.

In the SAR application, $P(\omega)$ is a unimodel function with peak at $\omega = 0$ so that P' is an odd function, zero at $\omega = 0$. This suggests writing the function

$$\frac{P'(\omega)}{[N_o + P(\omega)]^2} \equiv |h_1(\omega)|^2 - |h_2(\omega)|^2,$$

where

$$h_1(\omega) \equiv 0, \qquad \omega < 0,$$

$$h_2(\omega) \equiv 0, \qquad \omega > 0.$$

† Refer to Fig. 10.3 during the remaining discussion.

That is, h_1 represents a real, bandpass filter that passes only the part of the signal spectrum below ω_1, etc. Then, by Parseval's identity we have

$$I_{\alpha_1} = \int_{T_c} |S_t * h_1|^2 \, dt - \int_{T_c} |S_t * h_2|^2 \, dt. \qquad (10.29)$$

The calculation, Eq. (10.29) is accomplished by passing the real signal, on offset frequency $\omega_1 + \alpha_1$, through both filters h_1 and h_2, followed by a square-law device, and then summing, or integrating, for a time T_c. The integration can be approximated by a low-pass filter. Then the difference is formed which has a sign and is used to control the heterodyning oscillator. The whole circuit is shown in Fig. 10.3.

Of course, the error signal can be used to correct the center frequency by other means—e.g., controlling antenna or vehicle orientation.

Design considerations. Given a center frequency estimating and correcting feedback system of the general type derived and described, as shown in Fig. 10.3, the practical design is an interesting and somewhat involved problem.

Suppose the vehicle is a satellite: its pitch and yaw motions are, typically, crudely approximated by linear excursions, at roughly known maximum rates, between limit cycles. When a limit cycle is reached, a suitable combination of gas pulses reverses the rotational motion. Overall system considerations of weight, volume, vehicle life, etc., enter into the problem of how tightly the limit cycles can be set. Typically, these are greater than the beamwidth, and therefore the instantaneous vehicle position is unknown. Thus, the signal ensemble spectrum must be centered, and if we use the oscillator scheme, we have to bound the satellite motion rates so as not to introduce phase errors into the azimuth histories. Also, the order of the (linearized) system of Fig. 10.3 is important in determining what type of motions can be followed without serious lag errors in correction—a familiar control system design problem.

Such design would be accomplished using the linearized version of the system of Fig. 10.3. The derivation of this linear approximation and many design considerations have been considered in the context of SAR systems.

COMMENTS AND REFERENCES

The optimization of a linear, invariant processor under resolution and SNR criterion has been considered in the series of papers on phase errors by Brown and Palermo: see Brown and Palermo (1963a), Brown and Palermo (1965), and Brown (1967). The general situation has not been solved, perhaps because it is amenable only to numerical solution, as we found here.

The LMVE with additive noise and phase errors has been discussed by

Brown and Palermo (1963a), and in detail by Harger (1967a). The optimization of the complex modulation was discussed by Harger (1967a).

The discussion of the maximum likelihood criterion for a "known" signal with a phase error in additive noise was taken from Harger (1967b); there, the processor is discussed in some detail, and coherent optical systems that realize the processor are synthesized. The estimation of center frequency of a spectral density has been discussed by many people, as has been in particular the ML estimation—see, e.g., Levin (1965). An extensive treatment specifically for SAR systems has been given by Zelenka (1966) who includes the relevant bibliography.

References

Berglund, C. N. (1964). "A note on power-law devices and their effect on signal-to-noise ratio." *IEEE Trans. Inform. Theory* **IT-10**, 52–57.

Berkowitz, R. S., ed. (1965). "Modern Radar: Analysis, Evaluation, and System Design." Wiley, New York.

Brown, W. M., and Palermo, C. J. (1963a). "Theory of coherent systems." *IRE Trans. Mil. Electron.* **MIL-6**, 187–196.

Brown, W. M. (1963b). "Analysis of Linear Time-Invariant Systems." McGraw-Hill, New York.

Brown, W. M., Crane, R. B., Harger, R. O., and Palermo, C. J. (1964). "Theory and performance limitations of analog systems." *IEEE Trans. Mil. Electron.* **MIL-8**, 102–113.

Brown, W. M., and Palermo, C. J. (1965). "Effects of phase errors on resolution." *IEEE Trans. Mil. Electron.* **MIL-9**, 4–9.

Brown, W. M. (1967). "Synthetic aperture radar." *IEEE Trans. Aerospace Electron. Systems* **AES-3**, 217–229.

Cahn, C. R. (1961). "A note on signal-to-noise ratio in band-pass limiters." *IRE Trans. Inform. Theory* **IT-4**, 39–44.

Colwell, R. N. (1966). "Aerial photography—a valuable sensor for the scientist," *in* "Science in Progress," 15th Ser., W. R. Brode (ed). Yale Univ. Press, New Haven, Connecticut.

Cutrona, L. J., Leith, E. N., Palermo, C. J., and Porcello, L. J. (1960). "Optical data processing and filtering systems." *IRE Trans. Inform. Theory* **IT-6**, 386–400.

Cutrona, L. J., Vivian, W. E., Leith, E. N., and Hall, G. O. (1961). "A high-resolution radar combat-surveillance system." *IRE Trans. Mil. Electron.* **MIL-5**, 127–131.

Cutrona, L. J., and Hall, G. O. (1962). "A comparison of techniques for achieving fine azimuth resolution." *IRE Trans. Mil. Electron.* **MIL-6**, 119–133.

Cutrona, L. J., Leith, E. N., Porcello, L. J., and Vivian, W. E. (1966). "On the application of coherent optical processing techniques to synthetic aperture radar." *Proc. IEEE* **57**, 1026–1032.

Davenport, W. B., and Root, W. L. (1958). "Introduction to the Theory of Random Signals and Noise." McGraw-Hill, New York.

Elias, P. (1952). "Optics and communication theory." *Opt. Soc. Am.* **43**, 229–232.

Fowle, E. N. (1964). "A method of designing FM pulse compression signals." *IRE Trans. Inform. Theory* **IT-10**, 61–67.

Galejis, J. (1959). "Signal-to-noise ratios in smooth limiters," *IRE Trans. Inform. Theory* **5**, 79–85.

Greene, C. A., and Moller, R. T. (1962). "The effect of normally distributed random phase errors on synthetic array gain patterns." *IEEE Trans. Mil. Electron* **MIL-6**, 130–139.

Harger, R. O. (1963). "A note on the realization of ambiguity functions." *IRE Trans. Space Electron. Telemetry* **SET-9**, 127–130.

Harger, R. O. (1965a). "An analysis of recorder motion errors in optical processing." *Appl. Opt.* **4**, 383–386.

Harger, R. O. (1965b). "An optimum design of ambiguity function, antenna pattern, and signal for side-looking radars." *IEEE Trans. Mil. Electron.* **MIL-9**, 264–278.

Harger, R. O., and Crimmins, T. R. (1965). "The effect of phase errors on weighted spectra." *IEEE Trans. Mil. Electron.* **MIL-9**, 298–299.

Harger, R. O. (1967a). "Linear minimum variance estimation with complex phase errors." *IEEE Trans. Aerospace Electron. Systems* **AES-3**, 681–687.

Harger, R. O., and Lewinski, D. J. (1967). "Nonlinear effects of negative resistance amplifiers on signals." *Proc. IEEE* **55**, 1507.

Harger, R. O. (1967b). "On processing optical images propagated through the atmosphere." *IEEE Trans. Aerospace Electron. Systems* **AES-3**, 819–828.

Harger, R. O., and Crimmins, T. R. (1968). "The solution of a nonlinear design problem for synthetic aperture radars." *IEEE Trans. Aerospace Electron. Systems* **AES-4**, 316–317.

Heimiller, R. C. (1962). "Theory and evaluation of gain patterns of synthetic arrays." *IRE Trans. Mil. Electron.* **MIL-6**, 122–130.

Jones, J. J. (1963). "Hard-limiting of two signals in random noise." *IEEE Trans. Inform. Theory* **IT-9**, 34–43.

Kelly, E. J., Reed, I. S., and Root, W. L. (1960). "The detection of radar echoes in noise," I and II. *J. Soc. Ind. Appl. Math.* **8**, 309–341, and **8**, 481–507.

Kelly, D. H. (1960). "Systems analysis of the photographic process. I. Three-stage model." *J. Opt. Soc. Am.* **50**, 269–276.

Kerr, D. E. (1951). "Propagation of Short Radio Waves." McGraw-Hill, New York.

Klauder, J. R., Price, A. C., Darlington, S., and Alberscheim, W. J. (1960). "The theory and design of chirp radars." *Bell System Tech. J.* **39**, 745–808.

Leith, E. N., and Upatnieks, J. (1962). "Reconstructed wave-fronts and communicatio n theory." *J. Opt. Soc. Am.* **52**, 1123–1130.

Leith, E. N., and Upatnieks, J. (1965). "Photography by laser." *Sci. Am.* **212**, 24–35.

Levin, M. J. (1965). "Power spectrum parameter estimation." *IEEE Trans. Inform. Theory*, **11**, 100–106.

Magnus, G. W., and Oberhettinger, F. (1949). "Special Functions of Mathematical Physics," Chelsea, New York.

McCord, H. L. (1962). "The equivalence among three approaches to deriving synthetic array patterns and analyzing processing techniques." *IRE Trans. Mil. Electron.* **MIL-6**, 116–119.

Mentzer, J. R. (1955). "Scattering and Diffraction of Radio Waves." Pergamon Press, Oxford.

Muehleman, D. O., Goldstein, R., and Carpenter, R. (1965). "Review of radar astronomy." *IEEE Spectrum* **29**, Part I, 44–55, Part II, 78–89.

O'Neill, E. (1956). "Spatial filterings in optics." *IRE Trans. Inform. Theory* **IT-2**, 56–65.

Reisz, F., and Sz.-Nagy, B. (1955). "Functional Analysis," trans. from second French ed. by L. F. Boron. Ungar, New York.

Ridenour, L. N. (1947). "Radar System Engineering." McGraw-Hill, New York.

Root, W. L. (1962). "Radar resolution of closely spaced targets." *IRE Trans. Mil. Electron.* **MIL-6**, 197–203.

Siebert, S. McC. (1956). "A radar detection philosophy." *IRE Trans. Inform. Theory*, **IT-2**, 204–221.

Sherwin, C. W., Ruina, J. P., and Rawcliffe, R. D. (1962). "Some early developments in synthetic aperture radar systems." *IRE Trans. Mil. Electron.* **MIL-6**, 111–116.

Slepian, D., and Pollock, H. O. (1961). "Prolate spheroidal wave functions, Fourier analysis and uncertainty—I." *Bell Systems Tech. J.* **40**, 43–63.

Vanderlugt, A. (1964). "Signal detection by complex spatial filtering." *IEEE Trans. Inform. Theory* **IT-10**, 139–146.

Vanderlugt, A. (1966). "Operational notation for the analysis and synthesis of optical data processing systems." *Proc. IEEE*, **54**, 1055–1063.

Born, M. and Wolf, E., (1965). "Principles of Optics." Pergamon Press, New York.

Woodward, P. M. (1953). "Probability and Information Theory with Applications to Radar." McGraw-Hill, New York.

Zelenka, J. S. (1966). "The Optimization of a Frequency Tracker." Ph.D. thesis, University of Michigan, Ann Arbor, Michigan.

Subject Index

A

Additive noise, *see* Noise
Ambiguities, 69–70
Ambiguity function, 69–70, 172–214, *see also* Ambiguity problem
  definition, 171–172, 200
  optimum design of, 172–196
    antenna design, 176–185
    criteria, 173–176, 190–191, 194–195, 198–200
    example, 188–190
    signal design, 185–188, 192–194
  properties of, 200–207
Ambiguity problem, 197–214
  ambiguity avoidance, 208–210
  ambiguity suppression, 210–213
  defective solutions, 212–213
  fundamental parameter restraint, 2, 29
  range extension problem, 207–213
    definition, 207
Antenna, *see also* Ambiguity function
  far-field pattern, 60–61
  illumination function, 60
  near-field condition, 42
  optimum design for ambiguities, 176–185
Applications of SAR, 13–14
Azimuth channel, *see* Linear characterization of SAR
Azimuth history, *see* Azimuth modulation
Azimuth modulation, 37–43, *see also* Linear frequency modulation
  bandwidth, 26, 27, 38–39
  limiting resolution, 40
  spectrum, 55–58
  time bandwidth product, 40–42
Azimuth offset frequency, *see* Offset frequency

C

Chirp modulation, *see* Linear frequency modulation
Compression, *see* Signal, SAR received; Optical processor

D

Depression angle, 29
Digital processing, 9, 53, 55
  processing operation, 54
  real time requirement, 54–55
  storage requirement, 53
Dispersive errors, *see* Phase errors
Doppler modulation, 19–21, 30–37
Doppler space, 30–37

E

Errors, *see* various types

F

Fourier transform
  formation by optics, 115–117
    by propagation, 60–61

G

Geometrical optics, method of, 104–107
Geometry of SAR, 2, 226–227

H

Holography, 12–13, *see also* Offset frequency

I

Image
  formation with optics, 114–115

Image (*cont.*)
terrain
definition, 4
geometrical optics, prediction of, 104–107
physical optics, prediction of, 79–82
Integral, formation with optics, 117

L

Least-squares criterion, *see also* Optimum design, Phase errors
ambiguity problem, in, 198
no noise, with, 90–92
specific objects, for, 91–92
Limiter, 157–158
Linear characterization of SAR, 70–75
azimuth channel, 74–75
range channel, 71–74
Linear frequency modulation, 21, 43–49, 111
definition, 44
effects on
of motion errors, 153, *see also* Motion errors
of nonlinearities, 165–166
of phase errors, *see* Phase errors
matched filtering of, 46–49, 111, *see also* Optical processing
nonlinear frequency modulation, 185–188
sampling of, 50–52
Linear minimum variance estimation
with additive noise, 95–98
with nonlinearities, 166–170
with phase errors, 220–223

M

Maximum likelihood criterion
for additive noise, 98–102
for ambiguity problem, 198–199
Maximum likelihood estimates, 98–101
ambiguity problem, in, 199
Motion errors, 146–153
definition, 146
large bandwidth, 149–151
small bandwidth, 151–153
Motion, scatterer, 35–37
Multiplicative noise, *see* Phase errors

N

Noise, additive
optimum design for, 92–102
least-square criterion, 95–98
maximum likelihood, 98–102
with phase errors, *see* Phase errors
SNR criterion, 92–95
two-dimensional recorded, 103–104
Nonlinearities, 156–171
effects on
multiple signals, 160–166
linear F.M. signals, 165–166
random signals, 166–170
false images, 164–166
with low-pass signals, 169
optimum design with, 166–170
independence of phase of, 170
with secondary noise source, 166–170
threshold effect, 163–164
transform method, 156–160
weak signal suppression, 161–162

O

Offset frequency, 11–13, 27–28, 64, 226–227
choice of, 67
definitions, 64
estimation of, 225–231
optical processing, effect of in, 121–127
Optical processing, 109–128
elements of, 109–120
Kirchhoff–Huygens principle, 109–112
lens, operations of, 112–117
Fourier transforming, 115–117
imaging, 114–115
integrating, 117
simultaneous range and azimuth compression, 125
transmission functions, 117–120
Optical processor, 121–127
elementary, 9–13, 111
frequency domain form, 126–127
LIV filtering, 126–127
spatial domain form, 121–126

Optimum design
with additive noise, *see* Additive noise
with ambiguity criteria, *see* Ambiguity function
comparison with various criteria, 102, 103
with no noise, 86–92
least-square criterion, 90–92
resolution criterion, 86–90
with nonlinearities, 166–170
with phase errors, *see* Phase errors

P

Parameters, SAR system
nominal relation among, 1–2, 26–30, 70–75
optimum design of, *see* Ambiguity function, optimum design of
Phase errors, 129–146
bounds for system design, 145–146
cubic, 134–136
definition, 129
dispersive errors
definition, 129
effects, 145
effects on
ambiguity function, 144–145
LMVE system, 130, 141–144
matched-filter system, 130–141
large bandwidth, of, 137–141
small bandwidth, of, 132–137
resolution, 134, 136–140
signal-to-noise ratio, 130
example of, 131
linear, 132
optimum design for, under, 215–232
LMVE criterion, 220–223
signal modulation, 222–223
maximum likelihood criterion, 223–231
center frequency estimation and correction of, 225–231
resolution criterion, 215–219
SNR criterion, 219–220
quadratic, 133–134
sources of, 131
Physical optics, method of, 79–82
Pulse repetition frequence (PRF), *see* Sampling, frequency of

R

Range channel, *see* Linear characterization of SAR
Range offset frequency, *see* Offset frequency
Reflectivity density, 61–62, 79–82
physical optics, calculation by method of, 79–82
scattering cross-section, relation to, 76
Resolution
definitions of, 24, 39–40, 86–92
effects on
of nonlinearities, 165–166
of phase errors, 134, 136–140
nominal relation for, 24–25, 39–40, 42–43
optimum design criterion for
ambiguity, *see* Ambiguity function
no noise, 22, 86–92
phase errors, 215–219

S

Sampling, 1–2, 26, 28, 49–52, 64–65, 67–68
frequency of, nominal, 28, *see also* Ambiguity function, optimum design of
Scattered field, 61–62, *see also* Geometric optics, method of; Physical optics, method of
Scattering cross-section, 76
Sidelobes, *see* Resolution, ambiguity function
Signal, SAR received, 19–21, 60–66
bandwidth, 26
matched filtering, 21–26, 54, 68–70
optimum design under
ambiguity criterion, *see* Ambiguity function
least-squares criterion, 97–98
signal-to-noise ratio criterion, 94–95
spectrum, 66–68
Signal-to-noise ratio, 75–79
definitions, 76, 219
distributed objects, for, 78–79
optimum design criterion for
additive noise, 92–95
phase errors, 219–220

Signal-to-noise ratio (*cont.*)
simplified approximate derivation, 77–78
Stationary phase, method of
application, 18, 39, 48–49, 55–58, 192–194
explanation, 56–57

T

Terrain plane, definition, 63
Threshold effect, 163–164
Transform method for nonlinearities, 156–160

U

Unambiguous range interval, 28

V

Vertical, 29

W

Weak signal suppression, 161–162

ELECTRICAL SCIENCE

A Series of Monographs and Texts

Editors

Henry G. Booker
UNIVERSITY OF CALIFORNIA AT SAN DIEGO
LA JOLLA, CALIFORNIA

Nicholas DeClaris
UNIVERSITY OF MARYLAND
COLLEGE PARK, MARYLAND

Joseph E. Rowe. Nonlinear Electron-Wave Interaction Phenomena. 1965

Max J. O. Strutt. Semiconductor Devices: Volume I.
Semiconductors and Semiconductor Diodes. 1966

Austin Blaquiere. Nonlinear System Analysis. 1966

Victor Rumsey. Frequency Independent Antennas. 1966

Charles K. Birdsall and William B. Bridges. Electron Dynamics of Diode Regions. 1966

A. D. Kuz'min and A. E. Salomonovich. Radioastronomical Methods of Antenna Measurements. 1966

Charles Cook and Marvin Bernfeld. Radar Signals: An Introduction to Theory and Application. 1967

J. W. Crispin, Jr., and K. M. Siegel (eds.). Methods of Radar Cross Section Analysis. 1968

Giuseppe Biorci (ed.). Network and Switching Theory. 1968

Ernest C. Okress (ed.). Microwave Power Engineering:
Volume 1. Generation, Transmission, Rectification. 1968
Volume 2. Applications. 1968

T. R. Bashkow (ed.). Engineering Applications of Digital Computers. 1968

Julius T. Tou (ed.). Applied Automata Theory. 1968

Robert Lyon-Caen. Diodes, Transistors, and Integrated Circuits for Switching Systems. 1969

M. Ronald Wohlers. Lumped and Distributed Passive Networks. 1969

Michel Cuenod and Allen E. Durling. A Discrete-Time Approach for System Analysis. 1969

K. Kurokawa. An Introduction to the Theory of Microwave Circuits. 1969

H. K. Messerle. Energy Conversion Statics. 1969

George Tyras. Radiation and Propagation of Electromagnetic Waves. 1969

Georges Metzger and Jean-Paul Vabre. Transmission Lines with Pulse Excitation. 1969

C. L. Sheng. Threshold Logic. 1969

Dale M. Grimes. Electromagnetism and Quantum Theory. 1969

Robert O. Harger. Synthetic Aperture Radar Systems: Theory and Design. 1970

In Preparation

M. A. Lampert and P. Mark. Current Injection in Solids.

W. V. T. Rusch and P. D. Potter. Analysis of Reflector Antennas.